The Creative Wri

NEW WRITING VIEWPOINTS
Series Editor: Graeme Harper, *Oakland University, Rochester, USA*
Associate Editor: Dianne Donnelly, *University of South Florida, USA*

The overall aim of this series is to publish books which will ultimately inform teaching and research, but whose primary focus is on the analysis of creative writing practice and theory. There will also be books which deal directly with aspects of creative writing knowledge, with issues of genre, form and style, with the nature and experience of creativity, and with the learning of creative writing. They will all have in common a concern with excellence in application and in understanding, with creative writing practitioners and their work, and with informed analysis of creative writing as process as well as completed artefact.

All books in this series are externally peer-reviewed.

Full details of all the books in this series and of all our other publications can be found on http://www.multilingual-matters.com, or by writing to Multilingual Matters, St Nicholas House, 31-34 High Street, Bristol, BS1 2AW, UK.

NEW WRITING VIEWPOINTS: 18

The Creative Writer's Mind

Nigel Krauth

MULTILINGUAL MATTERS
Bristol • Jackson

DOI https://doi.org/10.21832/KRAUTH5355
Library of Congress Cataloging in Publication Data
A catalog record for this book is available from the Library of Congress.
Names: Krauth, Nigel, author.
Title: The Creative Writer's Mind/Nigel Krauth.
Description: Bristol; Jackson: Multilingual Matters, [2022] | Series: New
    Writing Viewpoints: 18 | Includes bibliographical references and index.
    | Summary: 'The Creative Writer's Mind is a book for creative writers:
    it sets out to cross the gap between creative writing and science,
    between the creative arts and cognitive research. It examines what
    cognitive psychology, neuroscience and literary studies can tell
    creative writers about the processes of their writing mind' – Provided
    by publisher.
Identifiers: LCCN 2021062310 (print) | LCCN 2021062311 (ebook) | ISBN
    9781800415355 (hardback) | ISBN 9781800415348 (paperback) | ISBN 9781800415379
    (epub) | ISBN 9781800415362 (pdf)
Subjects: LCSH: Authorship – Psychological aspects. | Creative writing. |
    Cognitive science.
Classification: LCC PN171.P83 K73 2022 (print) | LCC PN171.P83 (ebook) |
    DDC – dc23/eng/20220310
LC record available at https://lccn.loc.gov/2021062310
LC ebook record available at https://lccn.loc.gov/2021062311

British Library Cataloguing in Publication Data
A catalogue entry for this book is available from the British Library.

ISBN-13: 978-1-80041-535-5 (hbk)
ISBN-13: 978-1-80041-534-8 (pbk)

Multilingual Matters
UK: St Nicholas House, 31-34 High Street, Bristol, BS1 2AW, UK.
USA: Ingram, Jackson, TN, USA.

Website: www.multilingual-matters.com
Twitter: Multi_Ling_Mat
Facebook: https://www.facebook.com/multilingualmatters
Blog: www.channelviewpublications.wordpress.com

The policy of Multilingual Matters/Channel View Publications is to use papers that are
natural, renewable and recyclable products, made from wood grown in sustainable forests.
In the manufacturing process of our books, and to further support our policy, preference is
given to printers that have FSC and PEFC Chain of Custody certification. The FSC and/or
PEFC logos will appear on those books where full certification has been granted to the printer
concerned.

Typeset by Riverside Publishing Solutions Ltd.

Contents

Figures

Acknowledgments

My initial research for this book was undertaken for an article co-written with Chris Bowman, 'Ekphrasis and the writing process', published in 2017 in *New Writing: The International Journal for the Practice and Theory of Creative Writing* 15(1). Chapters 2 and 3 include some material adapted from ideas I contributed and words I wrote for that article. Any sections of the article originally written by Chris Bowman are referenced accordingly and reused with his permission. Thank you, Chris, for the great pleasure of our working together.

Chapter 4 includes material adapted from my article 'Fragmented narratives: Minding the textual gap', published in 2019 in *TEXT: Journal of Writing and Writing Courses* 23(2).

I thank the following colleagues who gave me ideas. In some discussions, we were clearly aware that I was writing this book; in others, unexpected clues found in seemingly unrelated matters sent me off on new paths of enquiry.

Charles Fernyhough and John Foxwell at Durham University hosted me at the Wellcome-funded Hearing the Voice Project Research Group and took the time to assist me in my early research. Thank you, Charles, for your invitation to join the proceedings of the project.

Todd Lubart agreed to meet me in Paris and made me – as a creative writer – feel very welcome in the realms of cognitive psychology.

I tried to meet Jane Piirto in the USA but I landed there in the holiday period. Just the same, I appreciated Jane's enthusiasm in our email and telephone conversations.

I thank my Head of School, Jamey Carson, for supporting this project throughout.

Others who helped me particularly are Ross Watkins, Sue Woolfe, Jono Lineen, Ira McGuire, Lachlan Jarrah, Valentina Maniacco, Hayley Scrivenor, Rohan Wilson and Kevan Manwaring.

Members of my family who gave me great ideas in weekend talk-fests are Jane Johnston and Jenny Allen.

Finally, I thank Graeme Harper and Anna Roderick for their continuing, excellent editorial support.

Citations

Photograph of fMRI writing experiment in progress (Figure I.1) © Martin Lotze. Used by permission of Martin Lotze.

Diagram of the Flower and Hayes writing process model from Flower & Hayes, 1981 (Figure 1.1). Used by permission of *College Composition and Communication* and publisher NCTE.

Krauth version of the Writing Process Wheel, 1991 (Figure 1.2) © the author.

Images of brain activity and writing processes from Erhard *et al.*, 2014 (Figure 1.3). Used by permission of Elsevier.

Excerpt from Raymond Federman, *Double or Nothing* (Figure 3.1) © 1998 by Raymond Federman. Used by permission of the University of Alabama Press.

Introduction

The Writer's Mind: What Happens when Writing Happens?

Here I set out on a significant adventure for a writer – an exploration of my own mind. Like any creative writer, I have used my mind extensively in producing the novels, short stories, plays, memoir pieces and poems that make up the body of my work. My mind provides me with the basic equipment for creative writing production: an internal screen on which I preview settings and action; an inner sound stage where I hear character voices; an internal whiteboard where I configure plot and character arcs; and a personal mixing board where I try out routines of word choice, rhythm, figurative language, dramatic effects, and the particular array of sensibilities that typify my work. Additionally, there is an archive workspace called memory where I access re-runs of settings, action, conversations, and the plethora of sensory events I have experienced previously in my life.

I do not claim that exploring my own mind is something new for a creative writer to do. Many through the ages have produced acclaimed poetry, fiction, drama, and exegetical writing by conscious investigation of their thinking and imagination. But I do claim, in the academic context of this study, that my adventure sets out to do something new in a sustained way: to study not only the products generated in this personal production studio, but also the studio itself. My plan is to describe, in language suited to creative writers' understanding, those crossing-points where the creative writing discipline correlates with the disciplines of cognitive psychology, neuroscience, and literary studies – and further, to do this with a methodology that *begins with* the view of the writer and their process. In other words, I want to deal with creativity science from the basic needs and tenets of people involved in the creative writing process – not to deal with creative writing from the basic needs and tenets of science. I acknowledge that a gallant group of creative writer/adventurers have set out on this academic journey before me, particularly Jane Piirto (1998, 2002), Sue Woolfe (2007), Kevin Brophy (2009), Maria Takolander (2015), Dianne Donnelly (2018), Julia Prendergast (2019, 2020) and the contributors to *TEXT* Special Issue 13 on creative writing cognition edited by Nigel McLoughlin and Donna Lee Brien (2012). They have provided excellent maps for the first part of my journey, but I hope to keep going into other regions, too.

A key question to address as I start is: What or where *is* the mind I am talking about? Clearly, we know where the creative writer's brain is situated, but research indicates that what we call *the mind* is located not only there in the head but is distributed elsewhere in the body, and perhaps extends beyond the body, too. While neurons perform fundamental activity inside the brain, they also exist throughout our bodies where they communicate with muscles, glands and skin; they report back about pain and well-being; and they fire up when we experience emotions. Creative writers are particularly aware of the body's contribution to the mind's workings: as a creative writer, my head continually negotiates with my body and what it tells me about both the present and the past (Krauth, 2010); I am fully aware that these signals are communications not only with my body's extremities but also with the world I am part of.

So, what goes on in my head when I write? How and why do I make the creative associations that produce my creative work? How do I relay my mind's experiences, and the interpreted experiences of others, into my writing? How does my mind negotiate with my body and the world? What can cognitive psychology, neuroscience, literary studies and previous research in creative writing tell me about these processes? How does knowing about my mind studio help me as a creative writer?

The writer's mind has been explored, conjectured about and argued over since Plato said a creative writer had to go 'out of his mind' in order to write (Plato, 2001: 41). This ancient notion caused a two-and-a-half-thousand-year debate over inspiration – where it comes from, and just how 'crazy' creative writers might (need to) be. There have been famous cases of great writers (including poets, novelists and dramatists) who were depressives, or suffered from bipolar disorder. Writers such as William Blake, Samuel Taylor Coleridge, Virginia Woolf, Tennessee Williams and Sylvia Plath are regularly included on the list of 'crazy creatives'. But, as Maria Takolander (2015: 2) says: 'There are a number of reasons to be suspicious of scholarship strongly correlating creativity with psychopathology'. She concludes: 'It is not madness but the bio-culturally embodied mind' – i.e. a mind fundamentally engaged with the process of living – 'that ultimately underlies poetic creativity' (2015: 13). Although '*I'm seeing things*' and '*I'm hearing voices*' are popular signifiers of mental health problems, there is professional agreement today that when writers write *they stay in their minds*, although they do see images in their heads and hear inner voices – interpretable as signs of madness, yes, but actually part of normal thinking. Writers don't just experience *visual* and *verbal* cognition, as almost everyone does, they put it to use. Seeing those still or moving images in the mind, hearing those narrator or character voices, is what writers do when they plan, compose, rehearse and execute their writing.

Creativity and the Writer's Process

Novelist Sue Woolfe's non-fiction book, *The Mystery of the Cleaning Lady* (2007), is, for me, the best available work on the topic of the creative writer's mind even though the work is now 15 years old. The book is known among creative writers and students in Australia, but less so farther afield. It started life as the exegesis for her novel, *The Secret Cure* (2003), written as part of a PhD submission at the University of Sydney. Its basic question is: What are we storytellers doing in our minds? She asked the question because in 1998, she says, 'I became so stranded in the writing of a novel, and cast around for help so wildly that I took to wondering whether neuroscience could rescue me' (Woolfe, 2007: 2). Woolfe's book is my starting point for the adventure into my own mind. It mapped more territory at the intersection between science and the creative writing process as writers know it – and is more understandable to writers – than any other book. After her years of research, however, Woolfe (2007: 87) admitted in the book that she became 'disappointed in the infant technology of neuroscience'. Nevertheless, her survey of creativity science taught her (and can teach her reader) a great deal about the creative mind's operation in the writing process.

In the chapter 'Loose and tight construing', Woolfe (2007: 87–101) gets to the nub of her argument: she finds in the work of creativity psychologists in the 1980s and 1990s key ideas for understanding thinking while writing. These are ideas about how people 'construe', how they deal with constructs in their minds. Psychologist, George Kelly, recognised 'tight' and 'loose' construing in mind-work. In tight construing – the usual mode of thinking in the western world – we focus our attention on problems *so successfully* that we do not allow much creativity to occur. In loose construing, however, we *defocus* – something our school teachers rarely encourage us to do – and in the process we allow in more creative possibilities.

Woolfe referred to psychologist Colin Martindale, who approached creativity as 'the novel and useful combination of concepts previously thought to be unrelated' (Vartanian *et al.*, 2009: 99). Martindale cited word-association experiments with small groups of subjects primed to respond quickly (in two different ways) to a word supplied by an examiner. They must either (a) focus on the word, or (b) defocus on it and allow it to take them to broader mind-associations (Martindale, 1995: 255–259).

For example, given the word *table*, a focused group came up with *chair*, *food*, *desk*, *top*, *cloth*, and *eat*, while a defocused group came up with the same words, but also said *ocean*, *victory*, *leg*.

Clearly, there are personal reasons for someone saying *ocean* in response to the word *table*, but this is seen as precisely the capability the creative person has – they can get their mind to think in a defocused way about a problem at hand:

> When attention is focused, a few nodes [in the brain's neural network] are highly activated. As attention is defocused, activation is more equitably spread out among a larger number of nodes. Since no nodes are extremely activated, there is less lateral inhibition. Thus nodes receiving small amounts of input are activated rather than being inhibited by strongly activated nodes. (Martindale, 1995: 256)

The Martindale experiments involved ordinary people responding to prompts given under experimental psychology conditions. In the case of the creative writer, however, this capability – to think dissociatively not solely when primed to do so – gets harnessed as a modus operandi.

Sue Woolfe said: 'creative people differ from others in that they can better tolerate ambivalence' (2007: 96). This is true of creative thinking whether it happens in the sciences or the creative arts. Creative writers actually go searching for the non-obvious, the non-conforming, the new. Romantic poet, John Keats, suggested the same notion 200 years ago when he talked about 'negative capability' – the ability to accept 'uncertainties, mysteries, doubts, without any irritable reaching after fact and reason' (Keats, 1966 [1818]: 329). The loose and tight construing that Woolfe evoked in *The Mystery of the Cleaning Lady* is related to the divergent and convergent thinking I will focus on in the following pages.

There are many branches of psychology – social, behavioural, physiological, educational, etc. – and creativity is a key topic in all of them. In this book, I will concentrate mainly on cognitive psychology and what it has to say about creative writing because from there, and from neuroscience, have come the most recent interesting ideas about the creative writer's mind.

Recent Science about Creative Writing Processes

In Chapter 9, 'Literary creativity', of *The Neuroscience of Creativity*, Anna Abraham (2018: 200–225) surveyed neuroscientific studies specifically aimed at the creative writing process. There were not many such studies up to 2018. Abraham identified a handful of brain-scan experiments, based on writing prose, undertaken in 2013 and 2014 (Abraham, 2018: 218–221), and two experiments focused on writing poetry and lyrics in 2012 and 2015 (2018: 221–222). Several more from the period 2004 to 2017, she found, looked at story generation and metaphor-making (2018: 216–218). From those studies she concluded that 'the brain networks of relevance in understanding literary creativity include the language network, default mode network, central executive network, and semantic cognition network' (2018: 224). She goes on:

> All in all, then, the pattern of findings in the creative writing studies, both prose and poetry/lyrics, shows consistency in that multiple brain networks are involved in idea generation in such contexts, and subcortical regions of the CEN [Central Executive Network], like the basal ganglia, are additionally recruited as a function of expertise. (Abraham, 2018: 222)

It seems we use both sides, and towards the front, of our brains when we are writing. In my own case, I find this is exactly the region I touch with my fingertips when I lean forward, elbows on desk, taking my head in my hands, trying to figure out what to do next when writing. Perhaps in holding my head like this I am appropriately giving my rostral lateral and dorsolateral prefrontal cortex a subtle massage. I have previously thought of it as a posture of imminent despair! Or, at least, of very hard thinking.

In each of the creative writing experiments she described, Abraham (2018) noted the time spent by the subjects on tasks. Because of the difficulties associated with the brain-scan apparatus, times during which the brain was monitored while writing, or thinking about writing, ranged from 60 second to 140 seconds (Abraham, 2018: 219, 221). Admittedly, a minute can be a long time in a writer's thinking, but also it can evaporate in no time at all. A minute can produce an exciting breakthrough – or nothing. Clearly a minute in a machine can give only a partial account. Aptly, Abraham (2018: 24) noted: 'It is necessary to consider the stages of creative writing from inception of idea to final product to gain a systemic perspective of the interplay between individual, environmental, and cultural factors'. As a neuroscientist, Abraham showed insight into the situation of the creative writer brought into the neuroscientific spotlight, and she made clear that there are many issues for further consideration.

Central to neuroscience is the EEG, PET and fMRI[1] equipment that provides the ability to scan neurocognitive activity during writing. Advances by cognitive studies in the last two decades have focused on reading and creativity, and on other processual activities associated with learning and decision-making. But the full focus of this research has yet to land on mature creative or professional writing processes. If anything, results in neuroscience have produced less certainty, rather than more, about what happens in our heads when we write. It needs to be noted, however, that current experimental conditions in neurocognitive research – for example, those in fMRI laboratories – are far removed from the normal conditions in which a writer writes. Figure I.1, taken during an fMRI experiment about creative writing (Erhard *et al.*, 2014), illustrates the problems inherent in scientifically capturing what goes on in the creative writer's head.

Laboratory conditions are designed to filter out non-significant interference. Figure I.1 shows what a laboratory needs to do to access the brain, which is a key step towards accessing the mind, but it does not convincingly depict a writer in touch with their process. Mostly it illustrates, to me, the idea that much of creative writing is *not* separate from interferences but, rather, actually engages with and utilises the interferences that go on around and inside us, the elements non-significant to science which the laboratory seeks to exclude. (For example, I might prefer a particular room to write in, or need a special cup on the desk in order to write happily, or I might need to hold an amulet such as a coin in my hand to link to a particular character.) Clearly there is no novelist, poet, dramatist or memoir writer who does their normal work – let alone their best work – lying supine with head rigidly fixed, looking at their words through a system of mirrors while manipulating a felt pen on paper regularly replaced by an assistant.

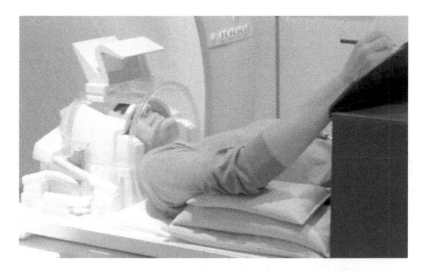

Figure I.1 fMRI writing experiment in progress (Photograph: Martin Lotze)

Erhard *et al.*'s (2014) pioneering study is titled: 'Professional training in creative writing is associated with enhanced fronto-striatal activity in a literary text continuation task.' It is the most widely publicised neuroscientific study of creative writing. It was reported in *The New York Times* in 2014 (Zimmer, 2014) and is cited in cognitive studies elsewhere, but not in the creative writing discipline. It monitored the brains of 20 '*expert writers*' versus 28 '*non-expert writers*' in a scanning machine (my italics). All subjects were students at German universities. The study focused on 'professional training' – so it was indeed about learning creative writing. But calling almost half these creative writing and journalism students 'experts' is problematic.

The participants undertook several tasks in the scanner. One was to brainstorm a story idea after two text prompts from contemporary German literature were given. They then had to write a new piece of work as an extension of those texts. Here is the result Erhard *et al.* came up with:

> Neuronal correlates of expert [sic] creative writing were associated with increased cognitive control located in prefrontal areas and with skill automatization processes involving the basal ganglia, in particular the left caudate nucleus [*thought to be responsible for effortless transfer of information and skill automatization due to its involvement in procedural memory and implicit motor control*]. In contrast, the increased activation of occipital areas in non-expert writers represented more visual and perceptual information processing. Altogether, our study is an important step towards understanding the neural processes underlying creative writing and expert literary education. (Erhard *et al.*, 2014: 22; my explanatory insertion in italics)

A basic understanding of the significance of these results – i.e. that writing becomes more automatic the more practice we get, and the more practised we are the less we need to visualise – suggests they are potentially *precisely opposite* to what is the case in reality. Thomas Mann (1992 [2003]: 397) famously said (and in doing so, represented the view of serious writers): 'a writer is somebody for whom writing is more difficult than it is for other people', which contradicts Erhard *et al.*'s (2014) first finding that writing becomes more automatic with experience. The exegetical perceptions of writers such as Italo Calvino, Joan Didion and Arthur Koestler indicate that control of *visualisation* while writing creatively is one of the most sophisticated skills possessed by mature, expert creative writers – certainly not a sign of immaturity as these results imply.

Eminent cognitive psychologist, Steven Pinker, criticised Erhard *et al.*'s study, saying: 'A better comparison would have been between writing a fictional story and writing an essay about some factual information' (Zimmer, 2014). Pinker pointed out that the brain activity seen by

Erhard during creative writing 'could be common to writing in general –
or perhaps to any kind of thinking that requires more focus than copying'.
He warned that:

> even the best-designed scanning experiments might miss signs of creativity...
> The very nature of creativity can make it different from one person to the
> next, and so it can be hard to see what different writers have in common...
> Marcel Proust might have activated the taste-perceiving regions of his brain
> when he recalled the flavor of a cookie. But another writer might rely more on
> sounds to evoke a time and place... Creativity is a perversely difficult thing to
> study. (Pinker, quoted in Zimmer, 2014)

Clearly more investigation is required.

Some recent activity focused on creative writing has taken place in
branches of psychology allied to neuroscience. The creative writing process
has attracted scientists using techniques from experimental cognitive
psychology (Fernyhough, 2016; Kaufman & Kaufman, 2009; Lubart, 2018),
while research into the mind-work required for creative writing appears
in studies of creative arts processes in general (Glăveanu, 2019; Kaufman
& Gregoire, 2015; Lubart, 2018). However, the available papers are more
understandable to psychologists than to creative writing researchers (as in
Kaufman & Kaufman, 2009). Research of interest to English departments,
which focuses on mind-activity in novels, tends not to see things from the
writer's processual viewpoint (Armstrong, 2013; Kemp, 2018; Palmer, 2004;
Tougaw, 2018). Mostly these works examine the literary product as a text
for readers, not as a text produced by writers; they interpret its production
in readerly ways according to literary archetypes and critical theory; they
are not especially interested in the synaptic detail of what goes on in the
writer's mind. Only one neurocognitive research paper I know of (Skov
et al., 2007) reviews in depth metacognitive autobiographical writing by a
master creative writer (Italo Calvino) as a point of entry into hypothesising
about writerly brain activity. In the following pages, where I refer to recent
science, I will focus mainly on work done by cognitive scientists, because
they have been the ones most interested in creative writing.

Research by Creative Writers

As mentioned above, a few academic creative writers, working
in the creative writing studies field, have focused on cognition from
the expert writer's point of view (Piirto, Woolfe, Brophy, etc.). The
number of individuals involved in this kind of research is surprisingly
limited – considering the centrality of brain activity to writing. Recent

how-to-write publications have applied some small measure of science to writing practice in popular contexts (Janzer, 2016; Reynolds, 2015). Dianne Donnelly's (2018) 'The convergence of creative writing processes and their neurological mapping' provides a serious survey of the neuroscience–creative writing interface. These works by creative writing scholars attempt to bridge the gap between disciplines, which is, of course, a challenge since each discipline speaks in its own specialised language. Needless to say, this book, *The Creative Writer's Mind*, attempts some gap-bridging too.

Creative writers have, in their own way, undertaken cognitive research. Jean-Jacques Rousseau, Mary Shelley, Arthur Koestler, Italo Calvino and Joan Didion, for example, described cognition in vivid exegetical essays. But most of the investigation has been done, and the results broadcast, embedded in creative writing itself, in the thinking of characters in fiction, poetry and plays. The *inner voicings* revealed in soliloquy or stream-of-consciousness narrative are the author's 'research results' from studying their own verbal cognition, or gained from what others reveal about their thinking. A character's description of *inner visualisation*, or the train of images they see in their head, expresses the kind of visual cognition the writer themself has experienced, or has learnt that others do. Authors are often asked by the public to reveal what goes on in their minds while writing, but they tend not to answer more than superficially. Perhaps it is because we realise these are the secrets of our trade, and we want to keep them to ourselves. What goes on in the studio stays in the studio, except in the packaged form – novel, poem, play, etc. – that we choose to let out the door. Perhaps it is because our creation of visual imagery and voice monologues or dialogues in the mind is so inherent to writing it goes unnoticed. In our normal lives we rarely stop to think about the way we think, and this includes the way we think while we are at work. In his 1986 poem 'Cosmopolitan Greetings', Allen Ginsberg (2003 [1986]) analysed the writing process. He advised writers to 'Catch yourself thinking' and gave reasons why it is useful to do so, among them being: 'Inside skull is vast as outside skull' and 'Mind is shapely, Art is shapely' (Ginsberg, 2003). Ginsberg drew our attention to the function, shape and contents of the mind in the writing process, particularly because we are so good at ignoring them, at taking them for granted, at observing outwards rather than inwards. This is good advice for any writer: *Look at what happens in your head when your writing happens.*

It's called meta-thinking, or metacognition. It is the process of paying attention to your own thinking. It involves thinking on two levels – doing your thinking while also 'stepping back' from it to analyse what thinking you are doing. For me, it's the process of internally viewing what's going on in my head. I don't do it normally; normally I just think. But I have found that I can think about things while at the same time think about how I am thinking about them. Not many writers have given us the benefit of a purview of their thinking while writing. One of the aims of this book is to bring to light more writers who have written about their meta-thinking, and to trace their processes.

Metacognition: Writers Thinking about Thinking

When we meta-think, we observe critically the images we see, the voices we hear, the smells, tastes and touching we experience in our head as conjured by the operation of our brain, and the effects of our senses on our brain. 'Metacognition is traditionally defined as the experience and knowledge we have about our own cognitive processes' (Schwartz & Perfect, 2002: 1); it involves '[t]hinking processes that permit control, selection and evaluation of one's own thinking' (Cropley, 2011: 435). We 'step back' from our immersion in thinking to observe it: to analyse it, deconstruct it, learn from it. In this process we feel we are being somehow objective about our subjectivity. We can note the processes our mind follows in thinking things through; we can understand, perhaps, *why* we understand and how or why we act upon those understandings. Admittedly, our minds are only in limited ways capable of understanding their own working, but metacognition has recently been applied in positive ways to learning skills, communication enhancement, eyewitness memory, emotional disorder therapies and sports training: 'metacognition has broad applications across a number of different settings ... metacognitive data from the lab have parallels to real-world phenomena and therefore can be applied' (Schwartz & Perfect, 2002: 1). In short, with metacognition we are being conscious of our own consciousness.

Studies of the role of metacognition in sport, for example, are relevant to creative writers. A paper focused on 'experts' metacognitive activities or their insights into, and regulation of, their own mental processes' argues that 'metacognitive processes and inferences play an important if neglected role in expertise' (MacIntyre *et al.*, 2014). This sounds like the situation in creative writing research, where study of the expert's mind has been limited. The MacIntyre paper concludes:

> Metacognition offers the potential to expand our understanding of expertise and individual domains of cognition through a rigorous examination of the mechanisms underlying self-initiated monitoring and control of one's own performance. (MacIntyre *et al.*, 2014)

Being aware of what goes on in the mind – tracing its workings, perceiving the steps being taken, following the direction of 'travel' and noting the quality of 'the journey' the mind takes in pursuing its objectives and in executing its accomplishments – is a useful tool with which to diagnose what might be effective, or what might be lacking, in our writing process. Meta-thinking provides opportunities to build an understanding of emergent skills, to maintain a flourishing practice, or to fix an ailing one. Writers who produced fiction, poetry and plays about the workings of the mind – from Sappho onwards – are clearly worthy of our attention because their study of *fictional* characters' minds allows us insight into how writers see cognition in action. It is fair to presume that these writers learnt a great deal about their characters' modes of thinking through experiencing, observing and studying metacognitively the workings of their own minds. And acclaimed writers such as Rousseau, Mary Shelley and Calvino have left us powerful accounts of the working of their minds in the context of their writing processes.

Rousseau introduced the modern 'individual self' into philosophy and literature. He investigated the idea of a unique identity 'distinguished from all social, cultural, and religious identities' (Coleman, 2008: vii) in his self-analysing, 600-page autobiographical project, *Confessions* (1782). It included investigations of his own mind – especially interesting for writers because he observed it at work while writing:

> It is with unbelievable difficulty that my ideas arrange themselves into any sort of order in my head. They circle there obscurely, they ferment to the point where they stir me, fire me, cause my heart to palpitate; and in the midst of all this emotion I see nothing clearly; I cannot write a word, I must wait. Imperceptibly, the great movement subsides, order succeeds chaos, everything finds its proper place; but slowly, and only after a long and confused agitation. (Rousseau, 2008 [1782]: 111)

Rousseau then developed a beautiful image of the mind as a big theatre stage:

> Have you ever been to the opera in Italy? While the scene is being changed in the great theatres there, an air of disorder prevails, which is disagreeable and lasts for quite a while: the sets are all muddled together; on every side there is a heaving and a pulling, which it is disturbing to watch; you are afraid it is all going to topple over. And yet little by little everything finds its place, nothing is missing, and you are astonished to see emerge from all this tumult a delightful spectacle. This process is more or less what goes on in my head when I am trying to write. (Rousseau, 2008: 111)

Rousseau continued his metacognitive analysis with a description of the writing output his disordered mind produced: 'Hence comes the extreme difficulty I have in writing. My manuscripts – crossed out, scribbled on, muddled, indecipherable – bear witness to what they have cost me.' Of course, Rousseau was recording problems that writers had before the computer and the delete button became available: 'There is not one of [my manuscripts] I have not had to copy out four or five times before giving it to the printer' (2008: 111). But his account of what goes on in the writer's head rings equally true for writers today. Processes in the mind have not changed in 250 years. (In later chapters, I will look at other writers' accounts of their minds in action when writing.)

Jumping forward from the 18th century to the 21st century, the concept of 'mindfulness' has entered the lexicon of creative writing studies with the publication of books such as Dinty Moore's *The Mindful Writer* (2012) and Joy Kenward's *The Joy of Mindful Writing* (2017). The writing approaches promoted by Moore and Kenward originate in Zen Buddhist meditation and include, among other techniques, the practitioner's focus on awareness of their thinking. 'Writing is an extension of thinking,' Kenward says. 'Thoughts will come into your mind... Notice the thought' (Kenward, 2017: 16, 18). Moore (2012) says: 'Mindfulness... is not limited to an awareness of your writing rituals and habits, or simply listening to the workings of your monkey mind'; it is also a practice which 'allows you to read your own writing, with preconceptions and attachments set momentarily aside. With proper focus, you can engage your own words as if they were written by someone else' (Moore, 2012: 73–74). The mindfulness qualities of openness, awareness and thinking newly, as defined by Ellen J. Langer, assist the writer in assessing their mind-work/writing by focusing on the process, not the outcome. On the subject of 'true process orientation' brought about by mindfulness, Langer says:

[It] also means being aware that every outcome is preceded by a process. Graduate students forget this all the time. They begin their dissertations with inordinate anxiety because they have seen other people's completed and polished work and mistakenly compare it to their own first tentative steps... they look in awe at Dr So-and-so's published book as if it had been born without effort or false starts, directly from brain to printed page. (Langer, 1989: 75–76)

According to cognitive scientists, there is a difference between *mindfulness* and *metacognition*, but the distinction is by no means clear. Tim van Gelder says:

Ellen Langer, the academic who brought the concept of mindfulness to prominence in social science, and more widely... says the key qualities of a mindful state of being [are]: (1) creation of new categories; (2) openness to new information; and (3) awareness of more than one perspective... Metacognition [on the other hand] is basically just thinking about one's own thinking, though the term generally also has the connotation that the thinking one is doing about one's thinking is aimed at or being used to improve that thinking... I can create new categories, be open to new information, and be aware of more than one perspective, without 'stepping back' and thinking about whether and how I am actually doing these things. (van Gelder, 2009)

Metacognition promotes itself as a means to understand and keep a check on how our learning and expertise happen. For writers interested in the mind's *modus operandi*, the approach encourages us to *think more about our thinking* in the writing process. Mindfulness publications offer possibilities for new writers keen to increase confidence in themselves and their processes, and provide insight into ways in which more mature creative writers might think differently about what they do. A writing practice does not mindlessly create product. Metacognitive skills monitor the studios we work in – those minds of ours where writing happens – and help us develop insight into the workings of other people's minds, too.

Where this Book is Heading

As a book for creative writers, *The Creative Writer's Mind* sets out from the creative writing side to cross the gap between writing and the creativity sciences, between the creative arts and neuroscientific/cognitive research. A creative writer might mind-image this book's challenging journey by seeing a wide chasm and a high-wire artist, balance pole swaying, taking tentative steps from the Creative Writing cliff-edge towards the Science cliff-edge on the other side of the gulf. The scientist, on the other hand, might mind-image the situation as a Venn diagram where the circles containing the sets Creative Writing (CW) and Science (S) do indeed engage, but the information in the CW ∩ S overlap is sparse, likely due to the fact that the languages spoken by the sets barely connect at all. So, this book is, for me as writer, a step into an unknown involving a difficult translating exercise – an attempt to find out if we as creative writers can have any sort of conversation with the natives in the *terra nullius* across the chasm.

In Chapter 1, I cite Heilman *et al.*'s article 'Creative innovation: Possible brain mechanisms' published in 2003. It is made up mainly of passages like this:

neurotransmitters such as norepinephrine might be important in CI. High levels of norepinephrine, produced by high rates of locus coeruleus firing, restrict the breadth of concept representations and increase the signal to noise ratio, but low levels of norepinephrine shift the brain toward intrinsic neuronal activation with an increase in the size of distributed concept representations and co-activation across modular networks. (Heilman *et al.*, 2003)

Helpfully, the authors of the article provide a 'translation' into language that the creative writing researcher can understand. But this assistance is not normal. Usually, creative writing scholars are left to their own devices in untangling the terminology, jargon and discipline perspectives when they try to make the interdisciplinary crossing. So, *The Creative Writer's Mind* sets out to be an interpreter by dealing with the writer's mind in terms that are native to the creative writing discipline. There already exist scientific and medical journalists who report to us valuably about what goes on in the sciences in language we can understand, and on occasions I have resorted to their writing in the popular press. But a joyous moment occurred for me when I realised that the most revered psychologist of all, the 'father of psychology' (and therefore a grandfather of neuroscience) and the first to offer a course in psychology at a university in the US, was William James, author of the classic text *The Principles of Psychology* (1890). Apart from coining the term *stream of consciousness* in his pioneering work, William was brother to Henry James, whose written works were also mind-study classics, published in the field of creative writing. Once I recognised *stream of consciousness* as the lynch-pin of my investigations, *The Creative Writer's Mind* was off and running. *Stream of consciousness* is a term deeply embedded in both cognitive science and creative writing, and the argument of this book subtends from that. In consequence, my project seeks to re-contact a lost history of creative writers writing about the mind, to develop that history further and to contrast it with current practitioner statements, and then contrast it again with what cognitive psychologists and neuroscientists say. In the process, I attempt to produce a cogent thesis about writers and their thinking when writing.

In her article, 'The convergence of creative writing processes and their neurological mapping', Dianne Donnelly (2018) overviews the creative writing process in the light of new developments in creativity studies. Donnelly brings to bear evidence provided by neuroscientists who mapped brain activity using fMRI scanners. She concludes that cross-disciplinary study of the creative writing process should continue because that would 'propel the growing field of creative writing research forward in America and... situate this knowledge within the larger creative writing domain in the US' (Donnelly, 2018: 96). The same should be said, I suggest, of creative writing research around the globe. Currently, academics who study creative writing – and creative processes more generally – do so in disciplinary silos. At the same time, cognitive researchers investigating creative writing rarely engage with recognised, successful writers or use them as subjects. Scientific research publications seldom cross-check with creative writing studies or literary studies research. Similarly, creative writing academics/practitioners almost never incorporate cognitive research into their work. Creative writing research is far more likely to turn to philosophy, sociology or cultural and literary studies to find its way forward – the implication being that disciplinary boundaries *within* the arts and humanities are more permeable than boundaries between the arts and the sciences.

While scholarly scientific articles with titles such as 'Essential skills for creative writing: Integrating multiple domain-specific perspectives' (Barbot *et al.*, 2012) cite dozens of scientific papers, not a single creative writer is referenced – not even an expert one who has talked exegetically about creative writing skills. (Other examples are Shah *et al.*, 2011; Erhard *et al.*, 2014 and Furst *et al.*, 2017.) Conversely, however, an exemplary article combining neuroscience and creative writing – Skov *et al.*'s (2007) 'Language and the brain's "mental cinema"' – begins by referencing the investigations of an expert writer: Italo Calvino. A key point here is that Skov *et al.*'s brain-scan research used Calvino's exegetical lecture, 'Visibility' in *Six Memos for the Next Millennium* (Calvino, 2016) as the *raison d'être* for undertaking the science. This acknowledges the fact that creative writers have laid foundations for other-discipline research in mind-activity.

An example of a scientific study of creative writing that exhibits little recourse to the expertise of creative writers is the *Psychology of Creative Writing* (Kaufman & Kaufman, 2009). This book includes 21 chapters and 31 authors, but only five authors publicise themselves as creative writers and another is recognised as a teacher of creative writing. On top of that, two of the authors are painters. The opening sentence of the preface is: 'Who is this strange being that is the creative writer?' (Kaufman & Kaufman, 2009: xix). It reminds me of the earliest explorers entering the *terra nullii* of Africa and Australia – ignorant of local knowledge, traditions and practices, yet claiming a superior right over the territory they penetrated. But I am also reminded that creative writing studies itself has not taken on the study of the mind. The *Psychology of Creative Writing* is a useful book because it traces so graphically the silo-isation of creative writing studies. It lets us, as creative writing practitioners and academics, know something of what is going on across the chasm. I do, however, hope that the following chapters shed greater light on the question of 'What goes on in writers' heads when they write?' as seen from the creative writing viewpoint.

Note

(1) EEG (Electroencephalography) uses electrodes placed on the scalp to measure neuronal electrical activity in the cerebral cortex. PET (Positron Emission Tomography) and fMRI (Functional Magnetic Resonance Imaging) both measure oxygenated blood flow changes associated with neuronal activity but they do it differently: PET traces a radioactive dye in the patient's bloodstream, while fMRI uses strong magnets, rather than radiation, to assess blood flow.

1 Depictions of the Creative Writing Mind

Interpreting the massive project of exploring the human mind by depicting character thinking and behaviour occupies many, many pages of literary criticism. There is not time here to canvass all that work. Essentially, the project was started in the West in ancient Greek drama and lyric poetry, and its development led – via many traditions, movements and genres of writing – through medieval and Elizabethan traditions of self-reflection, through Romantic meta-thinking and the emergence of the psychological novel, ultimately to filmscripts such as *Memento* and *Inception* and experimental multimodal works published electronically today. In all this creative writing, mental anguish over moral choice has been a key generator of the literary product; the mind-work of characters has driven action, plot and theme. To see how creative writers have dealt with the changing sensibilities and technologies available to them in the quest to describe the mind in action, we can look fleetingly at just one strand of the project: performance. In order to examine the mind in conflict, the ancient playwrights created a device – the Chorus – to evoke the workings of characters' minds and interpret them for the audience. By Shakespeare's time, the soliloquy was the preferred way to describe thinking on the stage. Today, technology can produce graphic depictions of the mind in the theatre and on film. Throughout the project's entire history, however, I dare say that the most efficient way for the mind-investigating writer to track decision-making and perception, has been in poetry and prose, particularly in creative works that utilise internal monologue and first-person narrative techniques.

But the widely acknowledged legitimacy of the creative writer as researcher of the mind lost traction over time as philosophers developed more systematic approaches. In the 19th century, modern psychology emerged to challenge philosophy of mind, and by the mid-20th century there was a specialised branch of science – cognitive psychology – given over to the experimental study of the mind. Nowadays, the influence of neuroscience impacts the field. While the mind has been portrayed in many ways – poetically, philosophically and scientifically – the following provides a brief history of depictions relevant to creative writing.

Muses

In the absence of better knowledge about how the mind worked, ancient creatives turned to the poetic concept of the Muse to explain inspiration and the creative process. Traditionally in western culture, nine Greek muses embodied the arts. Calliope was the muse of epic poetry, Clio the muse of history, Terpsichore the muse of dance, and so on. At the very beginning of western creative writing (although it was in the form of oral tradition back then) Homer called on muses to assist with his work. The first words of the *Iliad* (c. 800 BCE) – Αειδε θεὰ – are the exhortation, 'Sing, O Goddess' (Homer, 1888: 9), an address to a muse of song and creative composition, probably 'Calliope... the goddess of poetic inspiration' (Nagy, 2018), urging her to participate in the creation of the work. Following Homer, Greek poet Hesiod in his 'Hymn to the Muses' (c. 700 BCE) acknowledged: 'Happy is he whom the Muses love: sweet flows speech from his lips' (Hesiod, 2019). In the earliest oral story-telling and writing in western culture, it was accepted that creativity was muse-inspired, and that the makers of the best stories had minds that were divinity-assisted.

Arguments about the workings of the creative writer's mind started when Plato quoted Socrates in the *Phaedrus* and *Ion* dialogues (c. 370–390 BCE):

[A] poet is an airy thing, winged and holy, and he is not able to make poetry until he becomes inspired and goes out of his mind and his intel-lect is no longer with him. As long as a human being has his intellect in his possession he will always lack the power to make poetry... (Plato, 2001: 41)

This early theorising of creative production suggested that creative writing, seen as the dictation by a supernatural visitant to the mind of the writer, involved a necessary loss of rational control. In order to write creatively (as Plato's words indicate, without allowing them to be said with irony) the creative writer must feel possessed, hear voices, be unable to control the experience by intellect, and must copy down what the sweet voices say in their hijacked head. According to Plato in the *Phaedrus*, this was a madness like the madness of having prophetic visions, or being trans-ported by ritual drunkenness, or being crazily in love (Plato, n.d.: 113).

The dispute this theory caused, from Plato through to the 19th century, is traced by M.H. Abrams in his book, *The Mirror and the Lamp* (Abrams, 1971: 189–193). Here I pick out some of the highlights. As part of the debate, the astute Roman poet, Horace (65–8 BCE), went against Plato and saw art not as the product of madness but as 'a purposeful procedure, in which the end is foreseen from the beginning, part is fitted to part, and the whole is adapted to the anticipated effect upon the reader' (Abrams, 1971: 164). Shakespeare (1564–1616), a highly perceptive realist, too, called for a ramped-up 'Muse of fire' in his analysis of how writing for theatrical performance worked (*Henry V*, Prologue, c.1599): the 'purposeful procedure' he attributed to the muse was based in the effects the words of his scripts, combined with Elizabethan theatre architecture, had on the imaginative capabilities of his audiences' minds. In 1674, critic René Rapin (1621–1687) acknowledged the pervasiveness of the idea of the Muse in writing, and questioned the classical theory, but also had an each-way bet on it:

'Tis in no wise true, what most believe, That some little mixture of Madness goes to make up the character of a Poet; for though his Discourse ought in some manner to resemble that of one inspir'd: yet his mind must always be serene, that he may discern *when to let his Muse run mad*, and when to govern his Transports. (Rapin, 1674: 6; italics in original)

Romantic poet Percy Bysshe Shelley opted for the Muse idea in his 'A Defence of Poetry' (written in 1821). 'Poetry is indeed something divine,' he said, and questioned 'whether it be not an error to assert that the finest passages of poetry are produced by labour and study' (Shelley, 2001 [1840]: 713). Citing Milton's claim that the Muse 'dictated' to him the 'unpremeditated song' of *Paradise Lost*, Shelley likened writing to the plastic and pictorial arts:

Compositions so produced are to poetry what mosaic is to painting... a great statue or picture grows under the power of the artist as a child in the mother's womb, and the very mind which directs the hands in formation is incapable of accounting to itself for the origin, the gradations, or the media of the process. (Shelley, 2001: 714)

Shelley proposed that after an occasion of divine visitation, the writing mind then spends time translating the gifted message into communicable language. Thus the mind, inspired from outside, does an editing job. William Wordsworth had earlier suggested the same thing in his poem 'I Wandered Lonely as a Cloud' (1807) where he recounted that, after going for a walk, and while lying on his couch back home 'in pensive mood', the cosmic significance of the inspired moments when he saw the host of daffodils became clear to him, and at the same time became the subject of writing. Wordsworth analysed the writing process by saying that when he first saw the daffodils:

> I gazed – and gazed – but little thought
> What wealth the show to me had brought...
> (Wordsworth, 2002 [1807]: lines 17–18)

But then he completed his analysis of the process:

> For oft, when on my couch I lie
> In vacant or in pensive mood,
> They flash upon that inward eye
> Which is the bliss of solitude;
> And then my heart with pleasure fills,
> And dances with the daffodils.
> (Wordsworth 2002 [1807]: lines 19–24)

Wordsworth's study of the writer's mind is highly useful for creative writers. It teased out how the 'inward eye' of visual consciousness *re*views and *re*assesses the worldly experience of seeing. The poetic 'wealth' (line 18) of the experience is not necessarily fully recognised at the time of physical sighting but, with further mind-work done in solitude, significances for writing become clearer.

Wordsworth confirmed (in this poem and elsewhere in his oeuvre) that the role of the writer's mind was to process information gathered from the real world and turn it into writing, and he showed how he did it in practical terms – by walking and by lying down. A necessary part of that process, where the mind operates on its own with 'the bliss of solitude', involves thinking things through ('in pensive mood'), which engages cognitive visual re-enactment ('They flash upon that inward eye') and the re-creation of emotional experience ('then my heart... dances'). These mental inputs trigger the writing of the poem – its shape, movement and feel. For Wordsworth, the muse is Nature, and the dance with the muse is done while he rests on his couch in pre-writing and planning mode. In this poem, Wordsworth lays down the role of the creative writer's mind, which is: to find the significance of personal experience; to find its relationship to the rest of the world; and to find the structures whereby the experience might be delivered as writing.

In the 20th and 21st centuries, our understanding of how and why we write – and the mind's involvement in that process – continues to evoke the muse explanation, although the muse is now a down-to-earth concept among serious thinkers, having nothing to do with the divine. Of muses for men, Germaine Greer says:

> A muse is anything but a paid model. The muse in her purest aspect is the feminine part of the male artist, with which he must have intercourse if he is to bring into being a new work. She is the anima to his animus, the yin to his yang, except that, in a reversal of gender roles, she penetrates or inspires him and he gestates and brings forth, from the womb of the mind. (Greer, 2008: np)

There is a good dose of well-grounded irony in Greer's statement. Writers like William S. Burroughs and Ray Bradbury have also used the muse idea ironically; Burroughs said: 'Cheat your landlord if you can – and must – but do not try to short change the Muse' (Burroughs, 2012: 10). Bradbury, in a more developed personal analysis, said: 'The Muse… is that most terrified of all the virgins. She starts if she hears a sound, pales if you ask her questions, spins and vanishes if you disturb her dress' (Bradbury, 1996: 31). Bradbury continued:

> Another way of describing The Muse might be to reassess those little specks of light, those airy bubbles which float across everyone's vision, minute flaws in the lens or the outer, transparent skin of the eye. Unnoticed for years, when you first focus your attention on them, they can become unbearable nuisances, ruptures in one's attention at all hours of the day. They spoil what you are looking at, by getting in the way. People have gone to psychiatrists with the problem of "specks". The inevitable advice: ignore them, and they'll go away. The fact is, they don't go away; they remain, but we focus out beyond them, on the world and the world's ever-changing objects, as we should. So, too, with our Muse. If we focus beyond her, she regains her poise, and stands out of the way. (Bradbury, 1996: 32)

Bradbury's account suggests the muse is something to be avoided – seemingly a necessary, though poorly understood, part of the creative writing process, but essentially a hindrance. Female writers have taken a different tack, keen to reclaim a more intimate relationship between the creative writer and their process. Maya Angelou said: 'When I'm writing, I write. And then it's as if the muse is convinced that I'm serious and says, "Okay. Okay. I'll come"' (Angelou, quoted in Brunner, 2015). Isabel Allende (2013) advised: 'Show up, show up, show up, and after a while the muse shows up, too. If she doesn't show up invited, eventually she just shows up' (Allende, 2013: 6). I feel there is a lovely intimacy between women writers and their perceived muses, by which they mean the workings of their minds. It is very different from the idea that a male writer should become intimate with *her*, 'his muse'. As a male, I much prefer the idea that I become intimate with a part of *myself*, my own mind, in the creative process.

Clearly there have always been complications in the relationship between the writer and the concept of the muse. Today, much talk about the writing process goes on in the media and in popular conversation, which still reflects a mythologising of the creative writing process due to an absence of more convincing published knowledge. The advent of the technological and digital ages has not dislodged from popular belief the concept of the muse as responsible for creative practice, nor has it erased from serious writerly discourse the use of the muse-concept as metaphor. Most recently, a trademarked 'Muse' is available as 'a wearable brain-sensing headband. It measures your brain's activity using EEG (Electroencephalography) sensors' (Dodd, n.d.). You can buy it cheap on Amazon. The idea persists that a modicum of supernatural afflatus is involved in all creative mind processes.

Readers who appear to accept the muse explanation for creative writing do not, in my view, easily sense the experience the creative writer goes through. They examine the literary product *as text* and interpret its production in *readerly* ways according to societal and cultural expectation, media influence, normative education, literary archetypes and literary critical theory. If they are innocently amazed by the skills of masterful others, they reach for an easy 'Muse' explanation. Writers continue to fuel this convenient set of understandings. The muse explanation for creative work lends a mythic, quasi-historical quality to a reader's reading. Readers interested in reading *per se* are less likely to be interested in the actual cognitive or synaptic detail of what goes on in the writer's brain – they want something more dramatic and inspiring. After all, they fork out money to buy these expensive books! I think creative writing in the digital age is due for a proper demystification, even though it means wiping off some of that glamour, that fairy dust, which places the author on such a dubious pedestal.

Sciences

Until the 20th century, science allowed little status to studies of the mind and left it to philosophy and literature to deal with. While the mind could not be accessed and observed with methodological precision, non-scientific concepts of mind persisted. What was really needed for understanding creativity from the scientific viewpoint, was a breakthrough which allowed observation, with methodological certainty, of the working of the creative mind.

In the late 19th century, research in psychology made significant advances. William James, the recognised 'father of American psychology', proposed in his major work, *The Principles of Psychology* (1890), the concept of 'the Stream of Thought'. This involved, he claimed, his 'study of the mind from within' (James, W., 2019 [1890]: 224). His contemporary, the pioneering experimental psychologist Wilhelm Wundt, famously debunked James' ground-breaking book, saying: 'It is literature, it is beautiful, but it is not psychology' (Wundt, quoted in Steffens, 1931: 150). James undoubtedly undertook laboratory research – but he seems not to have been as masterful at it as was Wundt. However, James' individual thinking about stream of consciousness, and its associated depiction of convergent and divergent thinking, was highly influential in the early quarter of the 20th century, and it remains so today for scientists and writers alike. In 1880, delivering a lecture to the Harvard Natural History Society about 'the highest order of minds', James said:

> Instead of thoughts of concrete things patiently following one another in a beaten track of habitual suggestion, we have the most abrupt cross-cuts and transitions from one idea to another, the most rarefied abstractions and discriminations, the most unheard-of combinations of elements, the subtlest associations of analogy; in a word, we seem suddenly introduced into *a seething caldron of ideas*, where everything is fizzling and bobbing about in a state of bewildering activity, where partnerships can be joined or loosened in an instant, treadmill routine is unknown, and the unexpected seems the only law. According to the idiosyncrasy of the individual, the scintillations will have one character or another. They will be sallies of wit and humor; they will be flashes of poetry and eloquence; they will be constructions of dramatic fiction or of mechanical devices, logical or philosophic abstractions, business projects, or scientific hypotheses, with trains of experimental consequences based thereon; they will be musical sounds, or images of plastic beauty or picturesqueness, or visions of moral harmony. But, whatever their differences may be, they will all agree in this, – that their genesis is sudden and, as it were, spontaneous. That is to say, the same premises would not, in the mind of another individual, have engendered just that conclusion; although, when the conclusion is offered to the other individual, he may thoroughly accept and enjoy it, and envy the brilliancy of him to whom it first occurred. (James, W., 2009 [1880], italics added)

In their chapter in *The Cambridge Handbook of the Neuroscience of Creativity* (2018), Heilman and Fischler say:

> Creativity requires the novel understanding and expression of orderly relationships, and novelty requires that the creative person take a different direction from the prevailing modes of thought or expression, which is called divergent thinking. The concept of divergent thinking was put forth by William James [in 1890].... . Much of the empirical research on creativity since James' time has focused on divergent thinking as a critical element of creative innovation. (Heilman & Fischler, 2018: 477–479)

As a creative writer, I feel particularly at home with William James' view of what goes on in my mind, even though his metacognitive observations were written more than 130 years ago. It is notable that William James' younger brother was Henry James, a novelist who depicted the mind at work in terms similar to the 'seething caldron of ideas' used by William: or, as Henry put it, 'the great stewpot or crucible of the imagination' (James, H. 1972 [1911]: 77). One of William James' students at Harvard in the 1890s – Gertrude Stein – was another influential writer who focused on divergent thinking. I will come back to this group of ground-breakers later.

A key 20th-century breakthrough was made by British classicist and political scientist, Graham Wallas, in *The Art of Thought* (2014 [1926]). His book was an articulate and insightful discussion about the nature of thinking, and became, perhaps surprisingly, since Wallas was not science-trained, the go-to resource for creativity scientists even 90 years after its publication. I do not wish in any way to detract from Wallas's pioneering work by expressing surprise at its enduring influence. I do point out, however, that in an academic context any book with a title that includes the words *The Art of...* might be thought to stand no chance as a foundational science text in a century of laboratory experimentation in the cognitive and neuroscientific fields. As Teresa M. Amabile puts it:

> Despite being based almost exclusively on self-introspection and the introspective accounts of well-known creative individuals, the simple, plausible process that Wallas described has shaped a great deal of subsequent empirical research and theorizing in the psychology of creativity. (Amabile, 2019: 15)

Creative writing researchers – such as those published in *TEXT* and *New Writing*, and especially those who analyse practice – mainly ignore

Wallas's model, even though his categories are designed to identify major stages of practice universally applicable to the creative process. (For example, I can find only four references to Wallas in 24 years of *TEXT* publication.) Nevertheless, his 'venerable model' (Kozbelt, 2011: 475) is still recognised as fundamental to the scientific study of creativity (Amabile, 2019; Sadler-Smith, 2015).

In *The Art of Thought* Wallas identified four key stages in the creative process (Wallas, 2014: 37–39), which he later increased to five, as shown in my summary here:

(1) *preparation* (preparatory work on a problem that focuses the individual's mind on the problem and explores the problem's dimensions);

(2) *incubation* (where the problem is internalized into the unconscious mind and nothing appears externally to be happening);

(3) *intimation* (the creative person gets a 'feeling' that a solution is on its way);

(4) *illumination* or insight (where the creative idea bursts forth from its preconscious processing into conscious awareness); and

(5) *verification* (where the idea is consciously verified, elaborated, and then applied).

Wallas (2014: 39) characterised his stages of *preparation, incubation, intimation* (which he added later), *illumination* and *verification* as periods of 'conscious effort' – thus seeking to remove the Muse from the equation – although he explained that exploration of creative ideas 'may be unconsciously incubating' while other tasks of thinking were undertaken. In addition, Wallas described how '[i]n the daily stream of thought these [five] different stages *constantly overlap* each other as we explore different problems' (2014: 38; italics added). But when science uses the stages as the basis for experiments in creativity, it tends to ignore Wallas's insistence on 'overlapping' because, as five separable steps, the model is more elegant and scientifically useful. The Wallas model has been criticised as 'far too simplistic' (Kaufman & Gregoire, 2016: xviii) and I agree that, even when understood to comprise nuanced and variable stages, it is too blunt an instrument for describing the writing process (see Krauth, 2006, 2012). However, the Wallas stages provided a shared starting point from which scientists and creative artists alike could discuss and analyse creative thinking. As Rabovsky (2010: 7) said: 'The primary challenge for fruitful interdisciplinary work is to identify genuine contact points between disciplines'.

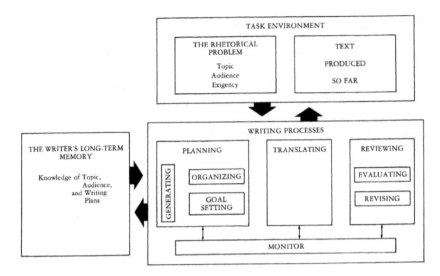

Figure 1.1 The Flower & Hayes writing process model (Flower & Hayes, 1981: 370)

In the 1970s and 1980s, the Wallas (2014) staged model of creativity was developed further by educators to provide resources for teaching composition to high school and college students. The teaching plans of the 1970s differentiated the stages of pre-writing, writing and re-writing in writing production, but treated the process as a one-way street, with no to-ing and fro-ing along the way. In 1981, Linda Flower and John R. Hayes published their cognitive process theory, which stated: 'The process of writing is best understood as a set of distinctive thinking processes which writers orchestrate or organize during the act of composing' (Flower & Hayes, 1981: 366). They produced a diagram to demonstrate the flow of mind-activities in the process (see Figure 1.1).

They provided a note on reading the diagram:

The arrows indicate that *information* flows from one box or process to another; that is, knowledge about the writing assignment or knowledge from memory can be transferred or used in the planning process, and information from planning can flow back the other way. What the arrows *do not mean* is that such information flows in a predictable left to right circuit, from one box to another as if the diagram were a one-way flow chart. This distinction is crucial because such a flow chart implies the very kind of stage model against which we wish to argue.

One of the central premises of the cognitive process theory presented here is that writers are constantly, instant by instant, orchestrating a battery of cognitive processes as they integrate planning, remembering, writing, and rereading. The multiple arrows, which are conventions in diagramming this sort of model, are unfortunately only weak indications of the complex and active organization of thinking processes which our work attempts to model. (Flower & Hayes, 1981: 386–387; italics in original)

Flower and Hayes noted:

The problem with [previously published] stage descriptions of writing is that they model the growth of the written product, not the inner process of the person producing it. [In them] 'Pre-Writing' is the stage before words emerge on paper; 'Writing' is the stage in which a product is being produced; and 'Re-Writing' is a final reworking of that product. Yet both common sense and research tell us that writers are constantly planning (pre-writing) and revising (re-writing) as they compose (write), not in clean-cut stages. Furthermore, the sharp distinctions stage models make between the operations of planning, writing, and revising may seriously distort how these activities work. For example... revision, as it is carried out by skilled writers, is not an end-of-the-line repair process, but is a constant process of 're-vision' or re-seeing that goes on while they are composing. A more accurate model of the composing process would need to recognize those basic thinking processes which unite planning and revision. Because stage models take the final product as their reference point, they offer an inadequate account of the more intimate, moment-by-moment intellectual process of composing. (Flower & Hayes, 1981: 367)

The Flower and Hayes model sought to overcome initial researcher misunderstanding of the Wallas model, where writing process was seen as linear, moving slickly from one stage to the next, all in a one-way flow. But the Flower and Hayes diagram, while helpful to educators, was too complex for use with students themselves in situations where they were learning to have insight into their creative writing processes.

By the early 1990s, educators had developed the Flower and Hayes (1981) diagram into the Writing Process Wheel figure. There were several versions of it available to teachers in books (there are many more available on the web nowadays). Figure 1.2 is a version I made myself in the 1990s as an overhead projection slide (before I had access to computer graphics – hence the cramped nature of my typed text):

Figure 1.2 Krauth version of the Writing Process Wheel, 1991 (© the author)

I used this primitive slide with first-year university creative writing students to explain to them what they did when they wrote. As a creative writer interested in describing to them what *I* actually did, I saw this diagram (of a warped sort of clock-face) as a major advance in my understanding of the writing process. The main point about my clock was that it drew attention to conscious thinking in the planning and pre-writing phases – called 'brainstorming' when done in groups in schools, but 'mind-work' when done individually – which was something I felt school education versions gave little attention to. A teacher cannot simply demand of a creative writing class: 'Write a story about this topic. Hit those keyboards now!' Similarly, a professional novelist does not think: 'I'll write a new novel. I'll start Chapter One today!' The earliest stages of writing – before any drafting occurs – involve conscious mind-work, and plenty of it, which may start days, months or years before the writer actually sets down a manuscript word on screen or on paper. (On my writing process clock, I told students that pre-writing took from noon to 5 o'clock, i.e. the gestation period in the mind was the most significant part of writing.) Also, my version of the wheel emphasised the re-involvement of earlier phases in much later phases in the process, i.e. reflecting Wallas's (2014) 'overlapping' and Flower and Hayes' (1981) two-way flow. I knew from my own writing practice that the process involved moving back and forth between stages, not only between adjacent stages but between all stages on the wheel.

A much better version of my Figure 1.2 diagram (which is lost some-where deep in my office files) shows double-pointed arrows crossing the gap in the middle of the wheel – producing a spoked wheel – indicating many 'going back and forth' tactics: for example, how a writer working on a re-write might recall something they thought of days, months or years previously in the pre-writing stage, and add it in. Or where an edit might conjure an unused/discarded enrichment stage idea, or even a rejected original inspiration idea, and find it useful for inclusion even at a very late stage. Apart from demonstrating to students an analysis of the thinking stages by which a creative work is produced, I also sought to impress upon them the agility of thinking required for the writing process.

It can be noted that the one big mystery on the writing process wheel is right at the top: 'Inspiration'. School education versions of the wheel did not expand on it, and neither did I, even though I often thought about its prominence. In class, it could lead to discussion of ideas about the age-old question asked of creatives: Where do you get your ideas from? But it remained an unexplained prompt at 12 noon *because* it was central to the debate that had gone on for millennia: What happens in the creative writer's mind? I could not define it, and serendipitously, it caused students to think about their own mind-activity.

Flower and Hayes acknowledged the experience of professional writers, and that is one reason why their model was a more realistic account of the writing process. In doing so, as education researchers they stole a march on creative writers themselves, who in the past had been notoriously unable to articulate well what they were thinking when they wrote. When asked about the secrets to writing, eminent writers had said things like: 'Make sure you have five sharpened pencils ready each morning', or 'Find a desk in a room which doesn't have a window to distract you; and make sure your chair-cushion is comfortable.' These were not the *secrets* to creative writing; they were superficial observations. As Flower and Hayes (1981) ultimately showed, the key processes affecting writing were cognitive, and particularly the flexibility and complexity of stages in thinking towards an outcome in a creative process.

Psychologists Frank Barron and Jane Piirto carried research-based study of the creative writer's mind through the 20th century. In a series of works from the 1950s to the 1990s, Barron researched imagination, creativity and the creative process in the arts and sciences. There was a wonderful moment in the history of the study of the writer's mind when Barron, in 1957, invited Beat poet Kenneth Rexroth to join a group of writers who would undertake psychological laboratory testing. Rexroth told the story this way:

> [A]n outfit calling itself the Institute of Personality Assessment and Research [IPAR], which had started out in life as the Office of Selection of the OSS, with the job of ensuring that everybody in the Army brightened the corner where he were, had got a potful of gold from the Carnegie people and had transferred from the Army to that River Rouge of the intellect, the University of California, and had cut loose on the creative personality, which they were very busy assessing and researching, and now they had got to writers and would I please come and be assessed? 'Our way of conducting research is to invite selected subjects to come to the Institute house for a period of two or three days; and there talk with members of our staff, participate in a series of experiments and psychological tests, and meet and interact with a number of other persons selected on a familiar basis'. (Rexroth, 1957: np)

Rexroth referred to the process as having his head 'tooken apart', and in reporting on the experience in the national press he did no favours for the idea that science might tell us more about how we write creatively. His conclusion was:

> What did it all mean? Nothing… Any Sioux medicine man, any kind and attentive priest, any properly aged grandmother, any Chinese herbalist, could have found out more in a half hour than these people did in three days…
>
> Theoretically this happy-go-lucky invasion of the sanctity of the person, this analysis of the sources of creativity and nonconformity, could be put to the most vicious uses. Not after some dictator has seized power – but simply in the due course of our obviously accelerating decay of liberty and respect for the person. (Rexroth, 1957: np)

Rexroth's reaction reflected not only widespread Cold War era scepticism about the usefulness of psychology, along with well-founded Beat paranoia about conservative society and government, but it also indicated the enduring sense that any analysis of the creative writing process by psychologists such as the eminent Frank Barron might somehow damage the mystique of how it all happened. Rexroth, however, went on to a stellar career as poet, essayist and translator, with no harm done by his brief stay answering questions and drinking cocktails at the Institute of Personality Assessment and Research.

It was not until the 1990s that a practising scientist who was also a creative writer provided a viewpoint informed from both sides. Jane Piirto, a creativity psychologist, poet and novelist, produced two ground-breaking books – *Understanding Those Who Create* (1998 [1992]) and '*My Teeming Brain*': *Understanding Creative Writers* (2002) – which brought together scientific enquiry, as psychologists and psychiatrists know it, with creative writing as creative writers know it. By focusing on the area where the two sets overlapped, Piirto identified rarely highlighted threads of research dating back through the 20th century and earlier. Piirto's books studied the *lives* of writers, their dispositions, childhoods and traumas; she adopted the perspectives taken by psychology but she did so in a manner specific to writers.

Piirto (2002: 3–21) observed how psychology spent time investigating creativity in the 20th century: how it coined the word *creativity*, which was not recorded in the *Oxford English Dictionary* until 1971 (Piirto, 2002: 7–8), and how it engaged with the idea of expertise, which is important when we attempt to analyse how the best creative writers do their writing (2002: 11–12). Piirto canvassed the history of overlapping enquiry in literature and psychology (2002: 13–16), for example showing how Freud, Jung, Lakoff and others incorporated the reading of literature into their scientific research to fuel their ideas about the Oedipus complex (Freud from Sophocles and Shakespeare), archetypal behaviours (Jung from Longfellow, the Gilgamesh epic, etc.) and the human mind's creation of metaphors (Lakoff and Turner (1989) from a variety of creative writers). In the context of looking at how storytelling underpinned much of what psychoanalysis and psychology investigated, Piirto researched biographical data from 160 successful and recognised writers (equally balanced between men and women) and used psychology as the framework to investigate their personalities, lives and behaviours, and the problems they experienced. Piirto found:

[C]reative writers were often early readers. They used early reading and writing to escape. They have often experienced childhood trauma, and may suffer from depression. They have high conceptual intelligence and high verbal intelligence. They are independent, nonconforming, and not interested in joining groups. They value self-expression and are productive. They are often driven, able to take rejection, and like to work alone for long periods of time. They often have difficulty with alcohol or substances. They prefer writing as their mode of expression of emotions and feelings. They often have advanced senses of humour. (Piirto, 2002: 146)

I expect there are writers who will say: 'Yes, that's me!' Others will disagree on various issues. The point was that Piirto (2002) focused on expert creative writers and attempted to categorise them psychologically according to actual data-gathering. Piirto identified 16 dominant themes in the lives of writers that affected their writing careers and she grouped these under five headings: home upbringing; community and culture; experience at school; chance (e.g. place of birth or ethnicity; and gender).

Piirto had a special interest in writers' psychologies but studied them biographically rather than experimentally. She was interested in what went on processually in their heads as they produced their work and she focused on the mental states and forces their living brought to their desks. She analysed aspects of creative processes highlighted by psychologists – including staged thinking (Wallas), flow thinking (Csikszentmihalyi), creative problem-solving (Isaksen, Puccio and Treffinger) and 'the process of a life' (Hillman, etc.). She also left space for other experiment-based creativity researchers to burrow into what actually went on in the head moment-by-moment when a writer engaged in writing (Piirto, 1998: 75–76; Piirto, 2011: 427–432). Piirto's studies used 'interviews, biographies and memoirs' (Piirto, 2018: 92) and may be less sexy than newly-exciting neuroimaging research, but the method of her investigations still underpins cognitive psychology and neuroscientific approaches today.

The Literary Imagination

I remember as a student and young literature lecturer in the 1970s discovering the usefulness of the term 'the literary imagination'. It worked as a catch-all to represent the complex set of mental manoeuvres the writer undertook in the creation of a literary work; it also justified them. There was something *sacred* or *alchemic* about the process whereby the author's mind turned the phenomena of reality, which everyone experienced, into the *higher form* of reality described in great literary works. It was a process by which the ordinary was infused with meaning; by which what we saw as normal around us took on greater significance. I still love this idea, many years after writing my own novels and teaching many students, but only an echo of the transcendental fervour I had for it exists now. I have today a more rational view of the dire need for the literary imagination in our world because, if anything, humans have become even more superficial in the way they see their surroundings and interpret their condition. We sorely need meaning to inform us in our lives, and the literary imagination continues to deal in that valuable commodity. That is not to say though, that in writing *this* book I wish in any way to devalue the beauty and excitement of the writing process, or to reduce it to routine. For me, back in the day, part of the usefulness – and mystique – of the term 'the literary imagination' was its *lack* of definition. Today, when I think about the literary imagination, I am inundated with a plethora of ideas about strategies, manoeuvres, techniques, and mind-tools employed by the writer in their process. It has not lost any of its mystique for me; it's as if I now know quite a lot about how alchemy happens, instead of being *totally* baffled by it. The mind was not physically observable back then. But science is catching up with philosophy and literature and can now give us insights into the mind-activities writers use: the visual images we see in our heads, and the voices we hear there, can be given more of 'a local habitation and a name' (as Shakespeare said in *A Midsummer Night's Dream* (2003–2021 [1595–1596]): V, i, 18) with scientific help.

The literary imagination has been the default literary studies term used to reference the mind of the creative writer as a kind of black box where divine assistance or magic happens, or at least as a mental workshop where the job of novelist, playwright or poet is done. J.J.A. Mooij's book, *Fictional Realities* (1993), traced the uses of the literary imagination, or writerly 'mental phenomena', from Plato and Aristotle until the middle of the 20th century, and pointed to 'the Romantic period as a culminating point, or hinge, of that history' (Mooij, 1993: 2). Mooij showed that the literary imagination had two major functions: creative and unifying. With the creative faculty, the writer saw the world and re-made it in their own mind, and the process of that re-making replicated the creative work of the divine Creator Himself (1993: 31–32). With the unifying faculty, the writer's mind used the imagination in connection with other mental faculties to procure 'a link between perception and understanding, or even between the highest flight of metaphysical inspiration and the common base of human knowledge' (Mooij, 1993: 32). Mooij reported also that the Romantics viewed the imagination as 'the supreme canceller and harmonizer of basic oppositions (like those between matter and mind, or object and subject)' (1993: 32). Thus, the literary imagination dealt not only with the world as it really was, but also with its ambiguity.

These days, in common parlance, *imagination* means something opposed to reality. But William Wordsworth's conviction, which 'bestowed on the word imagination a new meaning, almost entirely opposed to the ordinary one' (Legouis, 2000), strongly influenced the definition of the term *literary imagination*:

> He gave the name to his accurate, faithful and loving observation of nature. In his loftier moods, he used 'imagination' as a synonym of

'intuition,' of seeing into, and even through, reality, but he never admitted a divorce between it and reality... His hold over many thoughtful and, generally, mature minds is due to his having avowedly, and often, also, practically, made truth his primary object, beauty being only second. (Legouis, 2000: np)

So, the mind-work described as the literary imagination, although an 'essentially contested concept' in literary studies (Mooij, 1993: 1), deals with real issues and with finding instructive and entertaining ways to articulate the investigation of real human situations for consumption in novels, plays and poetry. I will not use the term *literary imagination* in this book, but I will be exploring the workshop and territory it refers to – the writer's mind – from the creative writer's own point of view.

Contemporary Depictions

The major shift for science in the 21st century has been to depict the writer's *brain* – rather than their *mind* – even while the two are spoken of now as the same thing. Hard science talks about *mind* as a physical entity – 'neural science is attempting to link molecules to mind' (Kandel *et al.*, 2000: 3), although, as Kandel said, 'consciousness poses fundamental problems for a biological theory of the mind... [but] despite philosophical cautions, neurobiologists have adopted a reductionist approach to consciousness' (Kandel, 2000: 396, 398). All the sciences, soft and hard, along with the arts, talk about mind, or consciousness, beyond electrochemical activity isolated in the brain: the 'embodied mind' is a universally discussed phenomenon. Some now interpret mind to encompass experience considerably farther afield – in the environment and culture (Glăveanu, 2018) – beyond the confines of the body.

The experimentally based disciplines cognitive psychology and neuroscience emerged in the 1950s and 1960s to focus on creativity in the general populace and the brain's engagement with problem-solving, education and language. Only some of this research targeted discrete creative processes among practitioners in branches of the arts, and little experimental focus landed on creative writing. Science turned occasional attention to the mind/brain of acclaimed artists/composers/musicians/dancers because those minds may tell us more about people generally than looking only at the creative mind-workings of the averagely talented. Science knows the brain is probably differently developed and configured for the differently gifted; therefore, looking at experts in any field may provide special understandings. But sadly, little brain-imaging has been done with creative writers.

According to contemporary science, our brains contain around 100 billion neurons with around 100 trillion connections between them. We develop these connections according to the mental and physical actions we undertake, the memories and abilities we acquire, and the skills we ultimately develop that make up our individual agency and identity in the world. These established capabilities are represented in the brain as *neural pathways*; for each of us our brain pathways dictate how we typically think and do; how we operate in thought when we approach a task or a topic; how we develop strategies for dealing with the everyday, the special or the new; and how our minds present our thinking to us in streams of words, images, memories, emotions, and plans for action. We develop these pathways simply by living and *learning* in our individual lives. But while experience shapes our brain and therefore our mind, we can improve and nuance these pathways with greater focused involvement or practice in body or mental activity, and with purposeful attention on maturing and progressing the understandings those pathways represent. Thus, if we practise writing well because we want to become a writer, we will likely establish better brain pathways significant to that ambition.

Heilman *et al.* (2003) described 'the neurobiological basis of creative innovation' – i.e. the chemical and electrical activity going on in the brains of creative people – in this way:

> CI [Creative Innovation] is defined as the ability to understand and express novel orderly relationships. A high level of general intelligence, domain specific knowledge and special skills are necessary components of creativity. Specialized knowledge is stored in specific portions of the temporal and parietal lobes. Some anatomic studies suggest that talented people might have alterations of specific regions of the posterior neocortical architecture, but further systematic studies are needed.
>
> Intelligence, knowledge and special skills, however, are not sufficient for CI. Developing alternative solutions or divergent thinking has been posited to be a critical element of CI, and clinical as well as functional imaging studies suggest that the frontal lobes are important for these activities. The frontal lobes have strong connections with the polymodal and supramodal regions of the temporal and parietal lobes where concepts and knowledge are stored. These connections might selectively inhibit and activate portions of posterior neocortex and thus be important for developing alternative solutions. Although extensive knowledge and divergent thinking together are critical for creativity they alone are insufficient for allowing a person to find the thread that unites. Finding this thread might require the binding of different forms

of knowledge, stored in separate cortical modules that have not been previously associated. Thus, CI might require the co-activation and communication between regions of the brain that ordinarily are not strongly connected...

Thus creative people may be endowed with brains that are capable of storing extensive specialized knowledge in their temporoparietal cortex, be capable of frontal mediated divergent thinking and have a special ability to modulate the frontal lobe-locus coeruleus (norepinephrine) system, such that during creative innovation cerebral levels of norepinephrine diminish, leading to the discovery of novel orderly relationships. (Heilman *et al.*, 2003)

Heilman *et al.* suggested that highly accomplished creative writers *might* have brains somewhat different from those of the general populace (I note the word *might* is used 47 times in the article). Required for creativity, it seems, are high levels of specialised knowledge (temporal and parietal lobes), capacity for divergent thinking (frontal lobe), and the ability to fire up that frontal lobe to get new pathways fizzing elsewhere in the brain. Notably also, the old idea of right-side brain creativity is not embraced here. In a 2005 article, neurologist Alice W. Flaherty suggested that the creative drive resulted from collaboration between the frontal lobes (which generate ideas), the temporal lobes (responsible for idea editing and evaluation), and dopamine from the limbic system (which 'plays important roles in executive function, motor control, motivation, arousal, reinforcement, and reward through signalling cascades' of feeling in the body) (Flaherty, 2005). Incidentally, Flaherty's findings also challenged the idea of right-brain creativity. As creative writers, we await agreement and certainty from the scientists.

Neuroscientific articles such as these suggest that the advanced artistic skill demonstrated in the kind of writing done by someone who has honed their craft over many years is not explained solely by looking at scientific experiments which focus on personality indicators, biographies, generalised problem-solving or aspects of low-level creativity in the general populace. What creative writing might ask for from scientists now is guidance as to what goes on in the mind and brain of established writers when they are in the process of actually producing work of excellent quality: How has 'excellence' been achieved? What has especially happened in the mind of the writer who produced a novel, poem, play, essay, etc., recognised as ground-breaking or bestselling? What processual thinking typifies the writer who produces exceptional work that changes a genre, a generation, a society, a publisher's bank account (or even their own)? Are such outcomes explained by saying that the author was simply the right person undertaking normal highly creative activity in the right place in the right language and culture at the right time? And seeing that individual expert writers will have individual neural pathways well developed, can we indeed trace patterns of exceptional thinking with the help of brain-scanning techniques that give us insight into the best ways practitioners might use their minds, and the best ways teachers can impart creative practice – so that aspiring writers stand a better chance of producing something of excellence, and so that cultures are more likely to benefit from individual creative endeavour? In 1933, Carl Jung said:

> We may expect psychological research, on the one hand, to explain the formation of a work of art, and on the other to reveal the factors that make a person artistically creative. (Jung, 1985 [1933]: 217)

But science has not yet delivered on Jung's 90-year-old statement because it is easier for science to generalise about whole populations as opposed to being specific about the mind-work of an individual in the creation of an individual work of art. Maybe what actually goes on in your head when you write that prize-winning or bestselling book or script will be a mystery for a long time yet.

Seventy years after Jung, neuroscientific studies with fMRI and similar scanning – for example, Erhard *et al.* (2014), Shah *et al.* (2011), Skov *et al.* (2007) – used neurocognitive technology to 'catch our thinking', as Ginsberg suggested, to help us better understand writing practice. We are now familiar with the type of visual depiction Erhard *et al.* made of creative writers' brains during writing (see Figure 1.3).

Creative Writing: expert writers

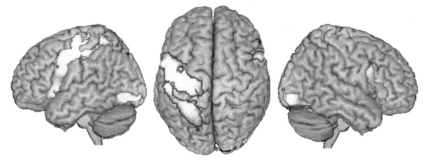

Figure 1.3 Brain activity and writing processes (adapted from Erhard *et al.*, 2014)

In the previous chapter, I gave my creative writer's reaction to Erhard *et al.*'s (2014) experiment, saying that the scientists' conclusions did not tally with a writer's conclusions. And it has to be said that neuroscience itself is uncertain about its basic tenets as brain visualisation shows more and more of the creative brain's workings. For example, long-held ideas about right- and left-side brain activity – the right supposedly being 'the exclusive font of creative thought' – have been challenged by the newer idea that *interaction between* right and left sides leads to greater creativity (Abraham, 2018: 81–85; Beaty *et al.*, 2016; Jacobs, 2017). Based on neuroimaging studies by David Dunson and Rex Jung, which looked at 'connections that span the hemispheres' and show that individuals with more interconnections tend to have higher creative reasoning scores, Jeffrey Kluger reports that:

> How a greater number of cross-hemisphere connections leads to greater creativity is unknown for now, but it's not hard to imagine that bringing more processing power to any problem – especially from parts of the brain that also bring different strengths and perspectives – could certainly lead to novel solutions. (Kluger, 2019: 14)

Our obvious inference at this point is to repeat Heilman *et al.*'s wise conclusion: *further systematic studies are needed.*

Wide-reaching accounts of neuroscientific research such as *The Neuroscience of Creativity* (Abraham, 2018) and *The Cambridge Handbook of the Neuroscience of Creativity* (Jung & Vartanian, 2018) indicate the extent of the science thus far. Anna Abraham (2018) has provided a summary of what neuroscience has achieved, by building on cognitive psychology, in studying creative writers' brains:

- Literary creativity is derived from the human capacity for language, which is inherently generative.
- Individual factors like intrinsic motivation and domain-specific expertise are essential to proficiency in literary creativity.
- It is necessary to consider the stages of creative writing from inception of idea to delivery of final product to gain a systematic perspective of the interplay between individual, environmental, and cultural factors.
- The brain networks of relevance in understanding literary creativity include the language network, default mode network, central executive network, and semantic cognition network.
- Paradigms based on verbal divergent thinking tasks are not necessarily implemented in a meaningful manner to deliver insights about verbal creativity.

- Neuroscientific studies are increasingly implementing creative writing paradigms that have the advantage of bearing higher ecological validity but also the disadvantage of being less controlled than the alternatives.
- There is an association between literary creativity and mental illness, but the precise nature and extent of the association is yet to be determined.

(Abraham, 2018: 224)

There are few surprises in Abraham's list depicting the creative writer's mind. Some points (e.g. 1, 2, 3 and 7) were to be expected, including the old chestnut that all writers are mad or substance abusers, etc. Some points, which indicate avenues towards new knowledge (e.g. 4 and 5) provide hope of more definitive information coming from brain-imaging. For my writing of this book, point 6 is most interesting because it admits the complexity of the task of describing what the writing brain does. It also acknowledges that usefully understanding the writer's process doesn't easily fit science's paradigms. Creative writing is a highly sophisticated, multifaceted, all-of-individual activity. To understand it, perhaps we need a *complete* knowledge of the mind and brain (and body) – something that is still a long way off. At the moment, the ultimate problem is that scientists can scan the brain but not the mind, even if we think of the two as the same thing. They can see the signs of thinking happening in the brain but cannot see the thinking itself. They can track how we think but not what we think. Until science comes up with a way for me to project my mind onto an auditorium screen for the rest of you to see and hear, then we are all in the dark, in our own private cinemas. (And do we want others to come and join us in here?)

In 2006, eight German universities collaboratively established The Berlin School of Mind and Brain. They said of their mission:

The application of neuroscientific methods to language research has become standard practice, involving such techniques as EEG/ERP/MEG to study its temporal dynamics, as well as fMRI studies to determine its spatial extent within the brain. However, there remains a substantial gap between the fine-grained linguistic theories of complex and diverse language phenomena on the one hand and current neuroscientific insights on the other. (Wartenburger & Spalek, 2010: 1)

What goes on with creative writing falls into this and other multidisciplinary gaps, not only because method, theory, terminology and research questions do not correlate, but also because 'the general way of thinking' not only about research but also about thinking itself, in science and arts fields is different (Bayer, 2010: 17). Researchers at the Berlin School say:

> ...very often interdisciplinary communication is at cross-purposes... results from one field often seem of little interest to another. The primary challenge for fruitful interdisciplinary work is to identify genuine contact points between disciplines. (Rabovsky, 2010: 7)

Neuroscientist Joseph LeDoux notes the changes occurring recently:

> Another way of looking at the humanist–science divide is to recognise that the old fault lines of academic debate are tumbling down. The turf wars that protect the sanctity of academic disciplines are giving way to cross-discipline interactions. Indeed, philosophers are not the only ones embracing neuroscience. Today neuro-fields are sprouting everywhere: neuroaesthetics, neuro-literary criticism, neuro economics, neuro-politics, neuro-marketing, and so on. (LeDoux, 2018: x)

To LeDoux's list can be added neuro-creative writing research. There's not much of it as yet, but there's a start.

Book-length studies, such as Kaufman and Kaufman's *The Psychology of Creative Writing* (2009), Kaufman and Gregoire's *Wired to Create: Unravelling the Mysteries of the Creative Mind* (2015) and Todd Lubart's *The Creative Process: Perspectives from Multiple Domains* (2018), operate effectively in the multidisciplinary area where creative writing and scientific enquiry overlap, bringing together neuroscience, cognitive psychology and creative writing practice along with practice in other creative arts areas. While it is early days yet to have confidence in pioneering neuro-imaging experiments involving writers, and while not all cognitive psychology findings necessarily convince creative writers, these books provide regularly useful insights. The situation for contemporary science's ability to depict the writer's mind is best summed up by Todd Lubart:

> There is probably no single 'creative process' that one can follow like a recipe to be sure to produce a creative product. Indeed, probably a multitude of paths can lead to a creative story (and an even greater number of paths can lead to a noncreative production). It may be possible, however, to identify the optimal process for a specific person to generate creative work given that individual's background and cognitive and personality profile, and taking into account that person's environment. (Lubart, 2009: 161)

In 2009, Lubart said: 'there remains much to explore about the nature of the creative process in writing' (2009: 161). Almost a decade later he noted encouragingly that: 'On the general topic of the creative process, there has been increasing attention in recent years' (Lubart, 2018: 2). In 2018, psychologist Jane Piirto said: 'A whole book could be written by this writer [herself] on the creative process in writers' (Piirto, 2018: 116), suggesting again that *there remains much to explore*. In the following chapters, I seek to take that exploration further.

2 Writers and Thinking, According to Critics

In his classic work, *The Modern Psychological Novel*, critic Leon Edel identified a historic change of direction in the modern novel at the end of the 19th century. This 'inward turning', Edel (1964 [1955]: 27) said, 'turned fiction away from external to internal reality, from the outer world that Balzac had charted a century before to the hidden world of fantasy and reverie into which there play constantly the life and perception of our senses' (1964: 11–12). In undertaking this inward turning, fiction writers became more involved in candid self-examination, and their writing became more *personal*, manifesting their need 'to cope with inner problems and project their inner life before the world'. The overall effect was that 'novelists sought to retain and record the "inwardness" of experience' (Edel, 1964: 12). The same effect occurred in other art forms: the visual arts turned to being about what our mind tells us is going on out there, not necessarily an 'objective' authority's point of view; music developed with highly personalised experimentation, related to inner emotion rather than societal paradigm; and poetry (which had regularly been about what the writer *thought*) set out on an even more confrontational, confessional and individualised trajectory. The inward turning, Edel noted (1964: 28), coincided with the rise of psychology and a scientific focus on the working of the mind, as publicised by Sigmund Freud's early work and by William James in his *Principles of Psychology* (1890).

William James' term *stream of consciousness* provides a contact point where literary studies, creative writing and science disciplines meet. While each discipline uses diverse terminology for mind phenomena, and each has given alternative designations to stream of consciousness itself, at an academic gathering of critics, writers and scientists, when 'stream of consciousness' is mentioned they all know what is being talked about. Ironically, we know also that 'William James the scientist was critical of his artistic junior [Henry James], who played with the facts of life in words instead of putting them under a microscope' (Edel, 1982: 125). The James brothers may have had their differences at the turn of the

20th century, but the study of the mind they advanced was ground-breaking for both psychology and creative writing. This chapter focuses on the roots of stream of consciousness in psychology, the ways it has been interpreted in literary studies to the present, and what creative writers might usefully gain from using the term as a tool in understanding their practice.

The Complexity of the Stream of Consciousness

The dominant modes of cognition are our verbal and visual thinking. While the mind processes other modes which engage our perception organs (touch, taste, smell) as well as our emotions, memories, bodily actions, etc., ultimately they all involve the stream of consciousness, which is the mind-activity described by psychologist William James in 1884 as 'the wonderful stream' we are aware of inside our heads (James, W., 1884: 2), and is more recently explained by creativity researcher Mark A. Runco as our 'inner psychic existence and functioning' (Runco, 2011b: 228).

Verbal thinking involves us using language in our minds. It is the wording, or voicing, that happens there: the monologue, the dialogue, the internal language experience (Fernyhough, 2010). While visual thinking (seeing images in our minds) also occurs, and cooperates significantly with verbal thinking in many situations, most of us think predominantly in language, or inner talk (Nęcka, 2011: 217–218). Scientists agree in principle that the two main ways we do our verbal thinking are *convergent* and *divergent* (Runco, 2011a; Russ & Dillon, 2011), analogous to writers recognising that mind-activity can be represented in a written work by *linear* sequential text or by *non-linear* fragmented text. Linear text reveals the mind working in logical, focused ways; non-linear text replicates mind-activity as less controlled and more messy. Writing in these different ways produces different sorts of narratives and narrative structures; different voices, plots and messages; and leads to recognition in different literary genres, either *conventional* or *experimental*. Scientists, correlatively, refer to convergent and divergent thinking and say that the former occurs where thinking operates to focus on an outcome, e.g. the solving of a problem, and leads logically or *convergently* towards an answer, while the latter explores creatively, draws possibilities in, and allows *divergent* associations to occur. Divergent thinking may be less targeted but it comes in handy for convergent problem-solving, too, e.g. in the form of brainstorming (Russ & Dillon, 2011: 70). It might be said that science generally favours thinking that leads directly to the pleasure of *a resolution* or *an answer*, i.e. science dislikes ambiguity and lack of resolution. But creative writing often favours thinking that leads to the pleasure of *further possibility*, e.g. the complicating of a plot, the uncertainty of suspense, the insight provided by metaphor, or the new awareness afforded by recognising the equivocal nature of existence. Lying behind (but not explaining) these attitudes to knowledge-seeking are popular notions that science, in its service to humanity, is a conservative, serious, credible endeavour to understand and explain reality. The arts, by contrast, are seen popularly as a radical, non-systematic and even sometimes frivolous endeavour to release us from reality's gravity.

In the light of the powerful division between convergent and divergent thinking, creative writers are faced with important questions: How should we react to represent thoughts in the head – as a streaming or as a series of separations? As a hierarchy or as a rhizome? As an arrow or a mosaic? As the way towards nailing down solutions or as the release into further thinking? That is, should writing operate by linear or non-linear means, given that thinking operates by both? And in dealing with a specific character's head, should the writer operate metacognitively, using their own understanding of thinking, or use cogent and smoothed-out logics provided from elsewhere – from religious dogma, folk psychology, political suasion, and so on? Other questions to ask might include: How best to describe a character's consciousness – with cliched thinking or with thinking unique to the character? What kind of thinking does the genre I write in expect me to use? How much should I foreground my character's psychology, as opposed to pushing it into the narrative background? Are my characters' minds important parts of the drama in this work? How am I actually creating my characters in this work?

In the late 19th century, while William James worked on his pioneering *Principles of Psychology*, brother Henry composed innovative psychological fiction including *The Portrait of a Lady* (1881) and 'The Aspern Papers' (1888). Each is remarkable for its trace of the to-ing and fro-ing of character thinking. The mind provides the main stage for the 'action' and 'setting' of these works; psychological drama is foregrounded as characters think about what they see and hear, what they should do and when they should do it, how others will react, the likelihood that their thinking represents the actual case in reality, and so on. Henry James' novels continue the tradition espoused by Shakespeare in Hamlet's soliloquy where not just a superficial description of thinking occurs, but an analysis of the nature of thinking is performed – its trajectories, its inconsistencies, its vicissitudes, the imaginings and memories and real events that influence it. Shakespeare and Henry James produced brilliant linear accounts of character thinking, with thought processes rationalised into flowing, dramatic verse and prose. These writers celebrated the logicality of the thinking voice while also acknowledging its potential breakdown in the face of harsh reality – when 'that noble and most sovereign reason' becomes 'jangled, out of tune and harsh', as Ophelia expressed it on seeing Hamlet's psychological deterioration (Shakespeare, 1963 [1600–1601]: III, i, 160–161).

William James coined the terms 'stream of thought' and 'stream of consciousness' in an 1884 article and popularised them in *The Principles of Psychology*. His pioneering work might be well over a century old now but it is still referenced in neuroscience (see, for example, McKiernan *et al.*, 2006). William James explored the two now-familiar depictions of thinking, linear and non-linear:

> Consciousness, then, does not appear to itself chopped up in bits. Such words as 'chain' or 'train' do not describe it fitly as it presents itself in the first instance. It is nothing jointed; it flows. A 'river' or a 'stream' are the metaphors by which it is most naturally described. (James, 2019: 239)

With his 'stream' and 'train' images, William James distinguished between the constant flow of thinking in the mind and the relatedness of the adjacent components that comprise the flow; he distinguished between *fluidity* of process (a given for the conscious mind) and *connectedness* of ideas (a variable). His contention is that we think all the time when we are conscious, and therefore thinking flows stream-like. On the other hand, the units of thinking that move along in this flow, he asserts, are segmented like train carriages or the links of a chain – they are separable into adjacent fragments and clusters that can be alike (in convergent thinking) or very different from each other (in divergent thinking):

> The transition between the thought of one object and the thought of another is no more a break in the *thought* than a joint in a bamboo is a break in the wood. It is a part of the *consciousness* as much as the joint is a part of the *bamboo*. (James, 2019: 240; italics in original)

The sections of bamboo, or carriages in the train, when alike and progressing sequentially, are recognised as *logical*. But when the segments of thinking follow each other in non-sequitur ways, we may have illogicality, but we may also have *creativity*. In his 1880 lecture to the Harvard Natural History Society, mentioned in Chapter 1 above, William James (2009 [1880]) described the thinking of 'the highest order of minds' as a flow which avoids 'thoughts of concrete things patiently following one another in a beaten track of habitual suggestion' and more often resembles 'a seething caldron of ideas, where everything is fizzling and bobbing about in a state of bewildering activity, where partnerships can be joined or loosened in an instant... and the unexpected seems the only law'. In effect he described graphically the kind of thinking that generated the revolutions in art and writing that emerged in Paris a quarter century later (as Cubism) and continued to develop worldwide in many forms of experimentalism during the 20th century. Significantly, James' student, Gertrude Stein, was at the heart of the French experimental modernist movements that foregrounded mosaic non-sequitur thinking as the basis for structure and narrative in artworks that reflected the workings of the mind. And notably, when Janes spoke of 'the highest order of minds' he was not selecting out geniuses, brilliant problem-solvers, and great creatives of the world: he was mainly distinguishing humans from 'the brutes... the dogs and horses', although he did admit that 'human intelligences of a simple order are very literal. They are slaves of habit, doing what they have been taught without variation' (James, W., 2009). James was, however, affirming that humans can think normally in ways that can also be highly creative.

Also of interest to writers, William James identified the varying pace of thinking flow and noticed how verbal and visual thoughts are interspersed:

> As we take... a general view of the wonderful stream of our consciousness, what strikes us first is this different pace of its parts. Like a bird's life, it seems to be made of an alternation of flights and perchings. The *rhythm of language* expresses this, where every thought is expressed in a sentence, and every sentence closed by a period. The resting-places are usually occupied by sensorial imaginations of some sort, whose peculiarity is that they can be held before the mind for an indefinite time, and contemplated without changing; the places of flight are filled with thoughts of relations, static or dynamic, that for the most part obtain between the matters contemplated in the periods of comparative rest. (James, 2019: 243; my italics)

William James specifically likens thinking to writing because both are fundamentally rhythmic. Writers talk of certain kinds of narrative as 'linear', since they incorporate 'train-carriage' couplings which *appear* to flow as James mentions, but are separable items comprising words and semantic clusters recognisable as phrases and sentences indicated by commas and full stops, etc. Truly linear narrative, one might argue, would be that which has no breaks in it at all. Early writing, as for example found in ancient Greek and Latin inscriptions and in medieval illustrated manuscripts, had no spaces between words because transcription tried at first to replicate the rhythmic, stream-like flow of speech (Ong, 2012 [1982]: 119). More recently, Gabriel García Márquez experimented with flowing prose in *The Autumn of the Patriarch* (1975), where a sentence or paragraph may last for pages (or seemingly endlessly). Many writers (e.g. Le Guin, 2004; Woolf, 2011–2021) have referred to the importance of rhythm in writing and how it is related to what happens in their minds or to their bodies. The point here is that William James, in introducing the idea of stream of consciousness to science, also prefigured foundational aspects of the discussion of linearity and non-linearity in creative writing.

When literary critics picked up from William James (2009, 2019) the idea of 'stream of consciousness', they went for the wrong term. For William James, *stream of consciousness* described merely the fact that we think continually while conscious. *Train of consciousness*, on the other hand, was the term he used to describe the variety of ways we put our thoughts together as segments to produce a flow. In this book, I will use William James' 'stream' and 'train' images to represent the two different modes of thinking he identified – continuous overall, and segmented within that continuity. It needs to be noted that *stream* and *train* thinking do not correspond to *convergent* and *divergent* modes of thinking; their relationship is not the contrast between 'zeroing in and expanding out' (Meeker, 1969: 20) which typifies the latter. Convergent and divergent thinking *both* occur within James' identification of 'train' or segmented thinking. The way they differ focuses on the differing relation between the adjacent segments: that is, does this train of thought involve carriages of a conceptually like and linked nature, or are they surprisingly and challengingly different?

The stream and train modes have preoccupied literary critics' descriptions of character-thinking since the early 20th century. James-*stream* thinking has dominated, but creative writers can apply James-*train* thinking to their processes, too. Like-segment, logically coupled thinking in the James-train is associated with writing in explicatory genres such as essay and scientific writing, due to its focused, convergent nature, but convergence is also required for the editing of a creative piece. James-train thinking with unlike couplings is useful in creative writing, especially at the planning and drafting stages – less so perhaps at editing stages – because fiction, poetry and drama are not only about ordered, rational thinking, they are also about depicting the variety and divergence of human response. Science writing is defined by its reflection of rational thinking and its aim to resolve matters of the physical world among the divergences of human and cultural endeavour. For science writing, James-train thinking involves like-segmented, or logically following carriages, even though initial inspiration for an investigation might involve very radical divergent thinking, and application of the findings in the write-up might require similar creativity. Scientists are told to follow the 'hourglass' technique: widely-applied divergent thinking first to create the project, convergent thinking to produce and analyse the results, and divergent thinking again for application of findings in the real world. James-train thinking for creative writing almost always involves steering away from what has been done before (except possibly in formularised genres) to identify new territories for exploration and new ways of putting words and sentences together to describe what is discovered there. Creative writing is defined by its reflection of, and making lively associations between, unresolved matters among the divergences of individual and cultural thinking. For creative writing, James-train divergent thinking is paramount, but James-train convergent thinking is also important because it helps produce a work a reader finds meaning in.

Critics on Creative Writing about Thinking

My topic here – 'Critics on creative writing about thinking' – is large and covers literary critical examination of narration, point of view, character creation, sense of self, memory, and other areas of creative writing involved with the representation of cognition. In the following discussion, I focus on stream of consciousness and how it has survived as a central concept in creative writing since William James coined the term in 1884.

In 1918, four years before Joyce's *Ulysses* was published as a monograph, literary critic and novelist May Sinclair was the first to apply William James' term *stream of consciousness* to a work of writing. She used it to describe the method in Dorothy Richardson's novel *Pointed Roofs* (1915). This led, it appears, to critic Elizabeth Drew in 1926 claiming that Richardson '"invented"' the technique and, in 1927, critic Katharine F. Gerould counterclaimed that Henry James '"introduced the method into English fiction"' (quoted in Bowling, 1950: 333). In 1950, Lawrence E. Bowling (1950: 333) picked up on the decades-old confusion among critics: 'The critics have failed to recognize different variations within the stream of consciousness technique,' he said. He pointed out that, because they did not distinguish between the voice of 'the whole of the consciousness' and a voice found 'nearest the unconscious' (1950: 334), their interpretation of what constituted thinking was unsound. Bowling's argument rested on the idea that *interior monologue* cannot include other than *verbal* elements, while *stream of consciousness* involves 'all conscious mental processes' including 'such non-language phenomena as images and sensations' (1950: 334). Bowling was right to some extent, and there is still plenty of confusion 70 years later. Bowling (1950) seems to have understood William James quite well, although he did not mention him. I would only beg to differ with Bowling on the question of whether interior monologue might include aspects of the other modes of thinking *by its reference to them*. When a character mentions in language something they see/hear/touch/taste/feel in their mind, mightn't we as readers experience it too, in our own way, as the character might do, via ekphrasis (see below).

The key problem facing empirical science is the difficulty of getting into minds to see what happens there. Outside science, in the everyday world where people collect, exchange and compare their thinking on a regular basis, personal accounts of mind occur repeatedly. So, too, in the literary world, investigation of consciousness occurs over and over in author descriptions of character thinking. In fiction, poetry and plays, it is standard practice for the writer to reveal what goes on in a character's mind and to examine how and why such thinking occurs. Creative writers analyse their own thinking, attempt to analyse how others think, transfer accounts of their own rational and imaginative thinking to fictional characters, produce arguments based on their own thinking which are designed to analyse the thinking process itself, and generally have mounted a campaign about thinking that today is described (or dismissed) as 'fiction' but might equally be seen as very like philosophy of mind in its metaphysical and subjective analysis, and not completely unlike cognitive psychology in making case studies of subjects' observed responses in physical and mental settings. Creative writers' thinking and writing about thinking has not had the same degree of impact as philosophy and cognitive sciences except perhaps with the stream of consciousness technique popularised by writers such as Henry James, James Joyce and Virginia Woolf.

Surprisingly little credence is given to the notion that creative writers have made a significant contribution to understanding the mind through their use of narrative devices including soliloquy, stream of consciousness, interior monologue and narratives of various kinds in first-, second- and third-person viewpoints. Major literary products that study a character's mind-work are acknowledged classics – Shakespeare's *Hamlet*, Henry James' *The Wings of the Dove*, Proust's *À la recherche du temps perdu*, Joyce's *Ulysses*, Woolf's *Mrs Dalloway* or *The Waves*, and many others. While a significant group of novels is now categorised as Psychological Fiction or Psychological Realism, authors across the board depict kinds of mind-activity in ways accordant to the genre they work in – from the detective's deductive thinking in crime fiction to the heroine's emotional thinking in romance fiction, and so on. According to Simon Kemp, 'every novel is a novel about the mind, at least to some extent' (Kemp, 2018: 3). But there is little special recognition that creative writers have explored and added importantly to our knowledge of mental processes.

Recent critical books that may change this perception include Simon Kemp's *Writing the Mind* (2018), Jason Tougaw's *The Elusive Brain* (2018), Alan Palmer's *Fictional Minds* (2004) and David Lodge's *Consciousness and the Novel* (2002). Earlier investigations of this territory were Dorrit Cohn's *Transparent Minds* (1978) and Sharon Cameron's *Thinking in Henry James* (1989). Cohn's book teased apart modes of introspective narrative used in fiction. It set up categories of monologue and discussed the relationship of *psycho-narration, quoted monologue* and *narrated monologue* (her terms) to memory, chronology, style, irony, etc. (Cohn, 1978: 11–14). Cohn compared critics' differing perspectives and she was sympathetic to the writer's viewpoint. She used William James (and other psychologists) as a touchstone for understanding the working of the mind. She pointed out that novelists such as Laurence Sterne:

> had used fragmentary syntax, staccato rhythms, non sequiturs and incongruous imagery when quoting minds in a state of agitation or reverie long before Jamesian, Freudian, Bergsonian, or Jungian ideas became fashionable. (Cohn, 1978: 84)

She included James Joyce among the 'great pioneer psychologists', William James and Freud, who had 'extraordinary powers of introspection' (1978: 97). In considering the techniques of 20th-century writers who used inner voices in their fiction – i.e. Woolf, Sarraute, Proust, Kafka, Joyce, Henry James, Faulkner and others – she created a valuable discussion of how consciousness was presented in first-person and third-person texts.

On the other hand, Sharon Cameron's book focused on how a single author (Henry James) during a lifetime of writing dealt with the challenge of representing in fiction the workings of consciousness. Henry James had developed the idea that a novel could be overtly and obsessively about the mind's versatility and vulnerability. But Cameron took Henry James to task for not producing a 'systematic' account of consciousness (Cameron, 1989: 35); she insisted that although he wrote his novels over a period of more than 25 years (1875–1902) – an era during which the orderly study of the mind (as done by William James, Freud and others) was in its infancy – he was culpable for revising or misrepresenting the attitudes and the techniques he used in his earlier works when he later wrote the 'Prefaces' (1907–1909). Referring to Henry James' 'investigations of consciousness' and 'revisions of [his] earlier conceptions of it', Cameron said:

> [I]n situating Henry James' revised conceptions of consciousness in the framework of his brother's (with which he was familiar) and of Freud's (with which probably he was not), I mean… to suggest a failed analogy between the scientific context and the literary one, for Henry James' revisions of consciousness are neither systematic nor signposted. Moreover, as they develop, they remain in conflict with each other. One could observe their instability by examining how consciousness is differently represented from one novel to another. (Cameron, 1989: 35–36)

In her 200-page book, Cameron mentions stream of consciousness just twice – briefly in the body of the text (1989: 34), then briefly in a footnote (1989: 186, n. 20). This is surprising, considering the centrality of the term to literary criticism. Although her book's title is *Thinking in Henry James*, this professor of English had no faith in the idea that a creative writer might make a serious study of the way thinking happens and, without needing to meet so-called scientific standards, present it in his novels. After William James and Sigmund Freud published their early major writings, psychology and psychiatry changed, rechanged, reworked, revised, deleted, chose to ignore, and also took much further, their theories about the mind. (Subsequently, of course, William James' and Freud's thinking ultimately metamorphosed into neuroscience.) While Cameron accused Henry James of not being 'systematic', she might more usefully have noted that, as a creative researcher into mind-activity, he produced a body of writing that spoke significantly to the explorations which science did at the time, and especially popularised the subject of those explorations – i.e. the complexity of the mind – among millions of readers in the world.

With the influence of cognitive psychology and neuroscience after the turn of the 21st century, studies in neuro-literary criticism emerged (see, for example, Armstrong, 2019; Hogan, 2014; Lesnik-Oberstein, 2017). Book-length works included Paul B. Armstrong's *How Literature Plays with the Brain: The Neuroscience of Reading and Art* (2013) and Norman N. Holland's *Literature and the Brain* (2009). I will focus, in the following, on studies which, apart from being about neuroscience and reading, also bring the writerly concept of stream of consciousness into their discussion.

In *Consciousness and the Novel* (2003 [2002]), novelist and critic David Lodge researched a wide range of 19th- and 20th-century writers in order to discover

> [h]ow the novel represents consciousness; how this contrasts with the way other narrative media, like film, represent it; how the consciousness, and the unconscious, of a creative writer do their work; how criticism can infer the nature of this process by formal analysis, or the creative writer by self-interrogation... (Lodge, 2003: xi)

Building on William James, and on his own earlier *The Art of Fiction* (1992), Lodge noted that around the turn of the 20th century 'reality was increasingly located in the private, subjective consciousness of individual selves' (Lodge, 1992: 42). He described interior monologue as the technique where 'we, as it were, overhear the character verbalizing his or her thoughts as they occur' (1992: 43). He said:

> ...it's rather like wearing headphones plugged into someone's brain, and monitoring an endless tape-recording of the subject's impressions, reflections, questions, memories and fantasies, as they are triggered either by physical sensations or the association of ideas. (Lodge, 1992: 47)

Lodge quoted Henry James' (ascribed) statement: that 'To project yourself into a consciousness of a person essentially your opposite requires the audacity of great genius; and even men of genius are cautious in approaching the problem' (Lodge, 2003: 50). He also identified parallels between Henry James' 1884 essay 'The art of fiction' and Virginia Woolf's 1919 essay 'Modern fiction', where both writers channel William James' ideas about consciousness. Henry James said:

> Experience is never limited and it is never complete; it is an immense sensibility, a kind of huge spider-web, of the finest silken threads, suspended in the chamber of consciousness, catching every air-borne particle in its tissue. (James, 2008: 5)

Virginia Woolf said:

> The mind receives a myriad impressions – trivial, fantastic, evanescent, or engraved with the sharpness of steel. From all sides they come, an incessant shower of innumerable atoms... life is a luminous halo, a semi-transparent envelope surrounding us from the beginning of consciousness to the end. (Woolf, 2009 [1919]: 9)

These statements can be compared with William James' 1880 lecture to the Harvard Natural History Society (James, 2009 [1880]) discussed above. Henry's essay was published in the same year that his brother published 'On some omissions of introspective psychology' (James, 1884) – an article in *Mind* journal where the original version of the stream of consciousness concept was launched – and just after publication of Henry's own early psychological novel, *The Portrait of a Lady* (1881), in which he developed to a significant level his techniques for entering characters' minds. Woolf's essay was later acknowledged as 'a manifesto for the modernist stream-of-consciousness novel' (Lodge, 2002: 51).

Lodge introduced into his argument not just the literary history of representing inner thought but also arguments made by a range of scientific thinkers, including cognitive scientist Daniel Dennett and neuroscientist Antonio Damasio. Lodge's contribution was summed up by reviewer Wingate Packard:

[Lodge's] project in *Consciousness and the Novel* is to affirm the value of the novel in describing or representing the human experience and to celebrate the literary intelligence that portrays human consciousness – the uniqueness each person feels when surveying the world from inside his or her own skull. What makes his writing so dynamic is that Lodge seems quite taken with the ideas about consciousness he finds in the fields of neuroscience, philosophy and cognitive and evolutionary psychology, but he is also confident that those fields have something to learn from literature... (Review by *Wingate Packard, The Seattle Times/Post-Intelligencer,* cited in *Reviews,* 2020)

Packard called Lodge's book 'finally courageous', as if to say, the comparing of creative writers' investigations with those of scientists' investigations is a risky business. At the same time, in *The Guardian*, philosopher and critic Galen Strawson called Lodge's book 'a pleasure to read as it makes its case for the novel's plausible claim to be the most powerful tool we have for the recording and examination of consciousness' (Strawson, 2002). Not full of praise, however, Strawson took both Lodge and William James to task for using the term *stream of consciousness* at all:

...it's not a very good metaphor. Streams have pools and falls, weeds and stones, not to mention waterboatmen and fish, and yet the suggestion of smooth, uninterrupted flow remains and is as inaccurate as Joyce's rendering of Stephen Dedalus's consciousness in *Ulysses* (1922) is accurate: 'Who watches me here? Who ever anywhere will read these written words? Signs on a white field. Somewhere to someone in your flutiest voice... .' (Strawson, 2002: np)

I think Strawson's rendering of a busy, interrupted (albeit overwhelmingly British) stream is perfectly appropriate to William James' and David Lodge's accounts of the mindstream. No worthwhile writer says our thinking is limpid, pristine and steady like an idyllic stream at the top of an unvisited mountain; it is, of course, chock-full of pollutions and disturbances.

Alan Palmer's *Fictional Minds* (2004) opened with a quote from Donna Leon's *A Sea of Troubles* (2001). Leon's novel is one of the popular Commissario Brunetti series set in Venice. In it the detective is talking to his wife, Paola, a literature lecturer at the university:

> 'We never know them well, do we?' [she says.]
> 'Who?'
> 'Real people.'
> 'What do you mean, "Real people"?'
> 'As opposed to people in books,' Paola explained. 'They're the only ones we ever really know well, or know truly... Maybe that's because they're the only ones about whom we get reliable information... Narrators never lie.'
> (Leon, 2002 [2001]: 318)

The pair are discussing how much they know of other people's minds, and how much they really know about each other's thoughts. It's a poignant moment, taking up the last paragraphs of the novel, a fitting conclusion to a work titled *A Sea of Troubles*, the phrase borrowed from Hamlet's famous soliloquy where he exposes his mind. But also, it is a fitting conclusion to this (or any) detective novel since the entire plot is based on the investigator's difficulties in penetrating the minds of murder suspects and reluctant witnesses, and the only evidence he has is what they choose to say, and what they give away about their real thinking through slip-ups of expression, gesture and action. Leon clearly intended Paola Brunetti's speech to be an ironic comment on her own [the author's] writing, but a powerful point is made about the impenetrability of others' minds for each and every one of us, and the degree to which we rely on fiction writers, for example, to provide us with at least some idea of how people think.

Fictional Minds is an analysis of how readers pick up and trace the mind-activity described in a novel. Early on, Palmer advocates not using the terms *stream of consciousness* or *interior monologue* at all, because of the confusion they have caused literary theorists in the past (Palmer, 2004: 23–25), although the terms remain as touchstones in his discussion throughout the book. He prefers to use *direct thought, thought report* and *free indirect thought* (2004: 54–55). His project is ultimately about 'the centrality of fictional minds to the reading process' (2004: 189) and his discussion explores not just fiction but also cognitive sciences and philosophy. He concludes:

> Fiction embodies what – in the fields of cognitive psychology, the philosophy of mind, and other cognitive sciences – is known as *folk psychology*. This label is intended to cover our standard, everyday, unthinking [sic], 'commonsense' assumptions about how our minds and the minds of others work. Fictional narrators employ folk psychology, and it would be unreasonable to expect novelists to do otherwise. (Palmer, 2004: 244; italics in original)

The folk-psychology view of consciousness in fiction taken by scientists is an oversimplification, as Palmer explains:

> The consciousness debate is concerned not only with folk-psychology notions of how we think but also with more counterintuitive versions of the process. An extreme example of this sort of theorizing is [Daniel] Dennett's argument in *Consciousness Explained* that there is no 'Cartesian theatre' in which a unified and single flow of consciousness takes place – what we experience as consciousness is merely an amalgam of the various 'multiple drafts' that are produced across all of the different regions of the whole brain. As soon as we become accustomed to this sensation, we experience it as a continuity of consciousness. Further thought needs to be given to whether or not there is a place for such obviously non-folk-psychological ideas in an analysis of fictional minds. For example, it might be that fictional-mind-construction in both the modernist and postmodern novel are, as Dennett himself, suggests, interestingly consistent with his ideas. (Palmer, 2004: 244–245)

Sometimes the prevarications of philosophers and scientists about what we experience in our minds can strike us as more fantastical than fiction itself. While science may hypothesise and philosophy conjecture, it seems to me, as a creative writer, that the inner voice I might give to a character – produced by and replicating my own actual, or a 'staged', inner voice – may indeed involve problems of authenticity and appropriation but is, in fact, a genuine reproduction of an inner voice, albeit my own. Perhaps the ultimate conundrum is that creative writing comes out of the author's stream of consciousness itself – as science findings come from the scientist's stream of consciousness, and philosophy from the philosopher's – so maybe we will never learn or know anything better.

In *The Elusive Brain: Literary Experiments in the Age of Neuroscience* (2018), Jason Tougaw says:

> While many works of literature can be described as direct responses to neuroscience, just as many represent the brain without engaging science directly. In that sense, it may be more accurate to describe them as responses to neuromania than to neuroscience. (Tougaw, 2018: 5)

Neuromania is the contemporary cultural tendency to exaggerate, over-simplify and distort recent developments in neuroscience (Tougaw, 2018: 4–5), but Tougaw embraces emergent works of brain memoir and neuron-ovel genres for the 'productive, instructive, and pleasurable energies' they generate (2018: 5). Neuroscientist, Joseph E. LeDoux, supports him in the foreword to the book: 'perhaps the humanities can help in ways that science may be limited', since science is particularly unclear about 'how first-person subjective experience comes about in the brain' (LeDoux, 2018: ix). LeDoux says:

> Tougaw's analysis of subjective experience in literature, especially in brain memoirs and neuronovels, provides useful insights into our experiences that may provide principles that can bridge humanist and scientific views and may help our [i.e. scientists'] questions about what to look for in the brain when we study subjective experiences. (LeDoux, 2018: x)

In a book that covers territory from autistic autobiographers to neuro-comics, Tougaw analyses a range of memoir and fiction in the light of neuroscientific experimentation. The creative works he looks at include Howard Dully's *My Lobotomy*, Alix Kates Shulman's *To Love What Is*, Siri Hustveldt's *The Shaking Woman, Or a History of My Nerves*, Thomas Harris's *Hannibal*, Ian McEwan's *Saturday*, Siri Hustveldt's *The Sorrows of an American*, John Wray's *Lowboy* and Maud Casey's *The Man Who Walked Away*, all published between 1999 and 2014. Tougaw keeps Jamesian theory in mind and is attracted to the idea of the thought-fringe at work in our thinking – the 'halo' around a thought whose affinities send us from one thought to another. This less-discussed aspect of William James' stream of consciousness is something creative writing scholars might link to the 'bowerbird researching' technique that Tess Brady describes as the 'haphazardness' of creative writing thinking (Brady, 2000). Tougaw says:

> In his essay 'The Stream of Consciousness', William James argues that 'the object before the mind always has a "Fringe"'. In other words, we are conscious of the objects we are paying attention to – [e.g.] the relations

between [our] brains, bodies, selves, and worlds – but our minds also vaguely sense meanings and associations of which we're not quite conscious... Through the process of putting words to page, we become conscious of what we didn't quite know beforehand. James argues that 'a good third of psychic life consists in these rapid premonitory perspective views of schemes of thought not yet articulate'. (Tougaw, 2018: 87–88)

The research thinking done by the creative writer, and the drafting and editing thinking that follows, involves recognising those associations in the 'fringe'. Tougaw calls this 'vaguely sens[ing] meanings and associations of which we're not quite conscious' and it happens in normal thinking. All writers become aware of it as part of their metacognitive stock in trade: creative writers depend on their brains to come up with new associations at the edge of thinking while they work, as do science writers, journalist, essay writers, etc.

Simon Kemp's *Writing the Mind: Representing Consciousness from Proust to the Present* (2018) begins by mentioning Marie Darrieussecq's *A Brief Stay with the Living*, published in 2001: 'a novel that draws on scientific research into consciousness for its themes, and on a hundred years of literary innovation in writing the mind for its style' (Kemp, 2018: 1). Kemp goes on to discuss seven experimental European writers (including Samuel Beckett) 'especially with a view to trace the representation of the mind through the century that separates' Proust from Darrieussecq (Kemp, 2018: 2). He asks:

Does 'mind' mean the same thing to all writers across a hundred years of cultural change and scientific progress? Do the fictional characters they invented, inhabiting fictional universes of their authors' creation, all possess identical mental architecture, or can we discern variations between them? If so, how do these various kinds of minds relate to our own minds, and to developments across the decades in the understanding of consciousness and the unconscious from outside the literary sphere? (Kemp, 2018: 2)

Kemp's *Writing the Mind* is a worthwhile book for creative writers to read. Unfortunately for English-speaking writers, it explores mainly non-English creative writers, perhaps implying that writers in English have not been experimental enough in grappling with the representation of the mind in fiction. For me, a key point that Kemp raises is the argument that Sartre made about Jamesian stream of consciousness. Sartre wrote:

In the end, it is not without a certain amount of doctoring that the flow of consciousness can be reduced to a succession of words, even malformed ones. If the word is given as an intermediary *signifying* a reality which, in its essence, transcends language, then so much the better: the word makes itself forgotten, offloading our attention onto the object. But if it presents itself as *mental reality*, if the author, in writing, claims to be giving us an ambiguous reality which is a sign in its objective essence, i.e. to the extent that it relates to something outside us, and a thing in its formal essence, i.e. as immediate mental datum in that case we can reproach him for not having taken a side, and for ignoring the rule of rhetoric which might be formulated thus: in literature, where you use signs, you must use *only* signs; and if the reality you wish to signify is a *word*, you must render it to the reader by means of different words. We might also reproach him for having forgotten that the greatest riches of mental life are *silent*. (Sartre, quoted in and translated by Kemp, 2018: 159; Sartre's italics)

It is hard to disagree with Sartre, who was not just a philosopher and critic but also an acclaimed novelist and playwright. Writing is 'doctoring' no matter how we deal with the voice in our head. The Surrealists tried to capture completely undoctored thought with automatic writing. Its results were dubious; it did not catch on with many readers, or indeed writers. But Surrealists were aware of silent – that is, visual and other non-linguistic – mental life and did much in the creative arts to engage it. Sartre's perceptive view of the writer's case, aware of what our minds can and can't do with words – as, too, with visuals, tastes, smells, emotions, etc. – suggests the complexity of the writer depicting the mind while also using the mind to depict it. Stream of consciousness is not a simple thing to observe metacognitively or to write down.

These studies remind us that presentation of consciousness has been a major preoccupation of fiction in modern times. The depiction of thinking is a key technique by which creative writers build character, justify action, shape plot, and construct the moral or philosophical message of a piece. Since the term *stream of consciousness* became widespread as part of humans understanding themselves, the mind has been a persistent focus for creative writers as a site of narrative viewpoint, a setting for psychological drama, and as a study for practitioners who know their success depends not only on understanding how readers' minds will respond to words set down but also on understanding what they themselves are doing in their writing process.

The Mind's Eye, the Inner Voice and Other Mind Processes Producing Writing

Scholarly writing about writing – as done in the academic creative writing studies field – refers often to critical studies beyond the creative writing discipline, i.e. in literary studies, philosophy, communications, etc.

An example occurs in the discussion of ekphrastic writing. In ekphrasis, the mind works to translate sense information from one mode (visual, auditory, haptic, olfactory, gustatory) into another. While ekphrasis has been discussed for a long time in rhetoric and in literary critical fields (Krauth & Bowman, 2017), scientists have recently provided us with greater clarity. Psychologist Edward Nęcka says:

> Humans, in opposition to other species, can perceive the external world through images or through words. In consequence, they can think and process information in two modes: verbal and nonverbal... [These two systems] based on separate codes of information processing... can cooperate, meaning that a piece of information obtained through the nonverbal channel can be verbalized, while verbal information can be expressed in mental images. (Nęcka, 2011: 217)

Nęcka confirms that viewing a visual object involves the creation of a visual image – a picture in the mind – which our nonvisual thinking can interpret and verbalise if we so wish, but also our thinking can turn that image into touch, taste or smell information, too. As part of the ordinary experience of living, we turn seen experience into verbalised experience (and vice versa) and we regularly verbalise or visualise other sensual experience as well. But just how we do this has been the subject of disagreement. In 1978, Roger N. Shepard wrote:

> Current controversy concerning mental imagery seems to have focused on two closely related questions: (a) Do the mental images that some of us undeniably experience play a significant functional role in our thinking or are they merely epiphenomenal accompaniments of underlying processes of a very different, less pictorial character? And (b) What exactly are mental images or, more specifically, what sort of physical processes underlie them in the brain, and to what extent are these processes, like pictures, isomorphic to the external objects they represent? (Shepard, 1978: 127)

To his credit as an eminent Stanford psychologist, Shepard, in this influential article 'The mental image' (published in *American Psychologist*), quoted Coleridge, Didion and other creative writers, as if to say that only poets and novelists up to that point had spoken sensibly about the creation of images in the mind and their usefulness in thinking. For example, Shepard says:

> The contemporary American novelist Joan Didion has asserted that none of the novels that she has written began with any notion of 'character' or 'plot' or even 'incident'. Instead, each developed out of what she calls

'pictures in my mind' – some of which she describes in vivid detail. Indeed, Didion claims that the very syntax and arrangement of the words of her sentences is dictated by the image: 'The arrangement you want can be found in the picture in your mind'. (Shepard, 1978: 127)

In 1978 there was debate among cognitive scientists (based on Galton's 19th-century findings) about whether images in the mind existed at all. Today it is apparent that some 2.5% of people do not see mind-images (the condition is called 'aphantasia', see Faw, 2009; Zeman *et al.*, 2015) but the argument has persisted (Pylyshyn, 2003). With the advent of brain-imaging techniques, scientists now suggest that the brain activity associated with internal visualising correlates with that of language production:

> Especially important has been the growing realisation that vision and language in some respects use a shared neural substrate... This research suggests, somewhat surprisingly, that imagery and semantic concepts may make use of the same neural representations, and may provide the key to explaining how 'image generation' and 'wording' are linked up in the creative process... (Skov *et al.*, 2007: 186)

Writing of every kind – from creative genres to scientific genres – requires an admixture of verbal and visual thinking, and oftentimes the translation of other senses into the verbal. As does living itself.

Looking specifically at the verbal processing mode, cognitive psychologist and novelist Charles Fernyhough provides an account of how we listen to the voice in our head as we think/compose/write, in his *The Voices Within: The History and Science of How We Talk to Ourselves* (2016). While Fernyhough (2016: 6) asserts that 'thinking is a multimedia experience' involving several modes at any one time (visual + verbal + tactile, etc.), he focuses on results from scientific experiments, which show:

> that the words that sound out in our heads play a vital part in our thinking. Psychologists are demonstrating that inner speech, as they term it, helps us to regulate our behaviour, motivate ourselves for action, evaluate those actions and even become conscious of our own selves. (Fernyhough, 2016: 11)

Writers familiar with stream of consciousness find psychologists' 'demonstrations' of it underwhelming. But Fernyhough is on-side with writers – he pays particular attention to the stream-of-consciousness technique in creative writing: 'The great modernist writers had a well-documented fascination with individual psychology and the artistic challenge of how to depict it on the page', he says. 'But writers were alert to the conversational quality of inner speech long before the era of Woolf and Joyce...' (Fernyhough, 2016: 93).

Fernyhough goes on to analyse Chaucer, Defoe and Charlotte Brontë as examples of writers who understood the *dialogic* nature of their characters' internal voices, and he acknowledges the debt his own studies have to literary theory, especially to Mikhail Bakhtin. By engaging with how the mind can create multiple voices to 'coordinate different perspectives' and achieve 'flexible articulation of different points of view' (2016: 99), Fernyhough provides insight for creative writers dealing with language in a heteroglossic world:

> One thing we can say about thinking is that it often appears to us as a kind of conversation between different voices propounding different points of view… Thinking is a dialogue… [When, as a child] you internalise dialogue, as you do when you develop inner speech, you internalise a structure that allows you to represent other perspectives. Those perspectives, in dialogic interplay, give your thinking some very special characteristics… the ability to coordinate different perspectives… and [bring] them into contact with each other. (Fernyhough, 2016: 12, 98–99)

This suggests we have a natural propensity to represent the voices and viewpoints of others; or, at least, writers who develop their sensibility for the dialogue of the voices in the head, can split their thinking apart, so to speak, and take on the voices of different (even opposed) characters. The ability to represent others' thinking and viewpoints is an especially desirable skill in these days of woke sensitivity to cultural, racial and gendered appropriation.

Alongside the case of producing character thinking in creative writing, the dialogic voice within is likely to be a writer's focus of attention when they write rhetorically or argumentatively – e.g. in essays, speeches and academic articles. We listen to and modulate the voice in our head as we develop the line of reasoning in a nonfiction work; we test the rhetorical logic of what we want to write *on an audience comprising ourself* in our mind. According to Fernyhough (2016: 99), the awareness that we think dialogically was 'there in the writings of Plato, William James, Charles Sanders Pierce, George Herbert Mead and Mikhail Bakhtin' but, oddly, it has 'never been spelled out in the terms of modern cognitive psychology'. At Fernyhough's Hearing the Voice project at Durham University, they continue to study the idea that 'inner speech is very common when people are writing' (2016: 186).

Fernyhough's argument is significant for creative writers. While much literary and philosophical theory was not 'spelled out' in terms of mind-activity, it was also distorted for creative writers by the way criticism developed. The literary history of ekphrasis provides an example (as discussed in Krauth & Bowman, 2017). Early uses of the term *ekphrasis*, as found in the Greek rhetoric handbooks (*progymnasmata*), allowed for a variety of definitions: 'a rhetorical exercise, a literary genre (or mode), a narrative digression, a species of description, or a poetic (even meta-poetic or meta-representational) technique' (Zeitlin, 2013: 17). Classics scholar, Froma Zeitlin, says: 'Only belatedly (or by implication) does ekphrasis include descriptions of works of art, the common usage today' (2013: 18). Originally, the standard topics for ekphrasis were not art works but 'battles, landscapes, festivals, seasons, people, animals, and object[s]' (Zeitlin, 2013: 18). The ekphrastic rhetoric exercise was about *describing* – the clear and vivid depicting/portraying/representing of an event, a view, a person or object in such a way that the 'enthrallment of the viewer' became the enthrallment of the 'would-be viewer' (2013: 30). The ekphrasis exercise sat alongside other exercises in the rhetoric hand-books, including those for narrative, personification and thesis. To excel at ekphrasis, the writer had to perfect the transfer of an image created *from reality* in their own mind, to the mind of another.

In the centuries following, the focus of the ekphrasis genre narrowed for writers until it was seen as a device mainly of relevance to specialist investi-gations of the relationship between visual art works and poetry. In the minds of most literary critics, this is the current usage. Three well-known critical works are examples – Emily Bilman's *Modern Ekphrasis* (2013), Stephen Cheeke's *Writing for Art: The Aesthetics of Ekphrasis* (2008) and James Heffernan's *Museum of Words: The Poetics of Ekphrasis from Homer to Ashbery* (2004 [1993]). They each look at ekphrasis as a process where a painting becomes a poem. Indicating the popularity of this view, Robert Denham's bibliographic tome, *Poets on Paintings* (2010), is a 200-page list of ekphrastic poetry volumes, anthologies, journals and websites plus an 80-page list of critical works to 2010. There are now more to add.

Having claimed a right to it – perhaps based on misplaced faith in the Plutarch (via Simonides) edict, 'Painting is silent poetry, and poetry is painting that speaks' (Plutarch, *De gloria Atheniensium*, 3.346f, quoted, for example, in Segal, 1995: 200) – the poetry industry does not tend to question whether its usage in any way limits the rhetoric handbooks' definitions of ekphrasis. Lately, however, other art forms have laid strong claims to the ekphrastic process. These include music, film and photog-raphy. Other writing forms have asserted themselves, too, including the short story, nonfiction and theatre.

We can focus, for example, on three critical works: Siglind Bruhn's *Musical Ekphrasis: Composers Responding to Poetry and Painting* (2000), Laura Eidt's *Writing and Filming the Painting: Ekphrasis in Literature and Film* (2008) and Andrew Miller's *Poetry, Photography, Ekphrasis: Lyrical Representations of Photographs from the 19th Century to the Present*

(2015). Each sets up a conversation between the 'traditional' participants in ekphrasis – poetry and painting – and a new area of involvement, i.e. music, film and photography. They call for 'expanding the definition of ekphrasis' (Eidt, 2008: 16) to include the creative processes they espouse. Bruhn and Eidt move the definition of ekphrasis forward to the idea where, ultimately, Eidt quotes W.J.T. Mitchell's classic *Picture Theory: Essays on Verbal and Visual Representation* (1994), saying that despite the ideological battles,

> there is no essential difference between the two arts [visual and textual]. In other words, neither are the visual arts 'inherently spatial, static, corporeal, and shapely', nor are 'arguments, addresses, ideas, and narratives' proper to language. Although the visual and verbal media are different 'at the level of sign-types, forms, materials of representation, and institutional traditions', Mitchell emphasizes that semantically, that is, in terms of 'expressing intentions and producing effects in a viewer/listener, there is no essential difference between texts and images'. Thus, in contrast to the restrictive, cautionary warnings of earlier critics, for Mitchell 'the problem is that we have not gone nearly far enough in our exploration of text-image relations'. (Eidt, 2008: 15)

Approaches to ekphrasis in the new millennium have opened up new attitudes, and new explorations question old beliefs that the mind worked in limited ways in creative writing practice. This reflects a new sense of intermediality – a greater awareness and understanding of the transfer of narratives between platforms in different media, between different writing and reading modes, and between different kinds of mind-activity. We know that we do not think in words alone anymore, nor in images alone: we do both together. The influence of screens in the computer age – and before that, film and TV – has changed our ways of reading and writing, giving them more spatial and less linear orientations (Kress, 2003, 2010). In the centuries following the establishment of the printing press, we grew dependent on the textual/verbal mode of reading and understanding. But in the computer age we are sophisticated ekphrasis participants: we have greater intermedial capability to turn images into words, and words into images – as also, ironically, humans were capable of before Gutenberg and printing, when oral and visual traditions prevailed and the cathedral with its exemplary murals, statuary, sculpted doors and stained-glass windows was as much a Bible for the illiterate listener as were the words read out by the literate priest (see Calvino, 2016: 105–106; McGann, 2001: xii). The process where a writer translates a painting into words today no longer holds the status it held for the last 600 years. The parameters for the practice of ekphrasis have changed.

Zeitlin refers to this when she says:

> ...beyond this brief definition [i.e. *enargeia* (vividness), *sapheneia* (clarity), and *phantasia* (mental image)], the word 'ekphrasis' immediately ushers us into a whole set of questions regarding its intermedial status in a potential contest between verbal and visual representations, the uses of mimesis with regard to verisimilitude (reality–illusion; truth–fiction), and its cognitive, psychological, and mnemonic values in the cultural expectations of its era. It would not be hyperbole to suggest that no other rhetorical term has aroused such interest in recent years among classicists and non-classicists alike, involving aesthetic considerations, theories of vision, modes of viewing, mental impressions, and the complex relationships between word and image. (Zeitlin, 2013: 17)

Traditionally, for creative writers, ekphrasis is an exchange between the seen and the written where the experience of a perceived visual object is transcribed into a verbal/written account, with special attention paid to the idea that the reader 're-experience' the original sensory encounter. Critical studies beyond literature show that ekphrasis involves exchange between *all* sensory input modes – visual, auditory, touch, taste and smell – to produce writing and other creative products. From the cognitive scientists' viewpoint, the case of the person undertaking an ekphrastic exercise to produce verbalisation/writing involves a transfer from one mode of neurological coding to another – from a mind-image, or a mind-feel, or a mind-taste, or a mind-smell, etc., to words expressed initially by the inner voice. In the case of the person hearing or reading the ekphrastic account, the transfer goes in reverse, from word to image or to other mind-experience. The ekphrastic process draws the writer's attention to the fact that they write with verbal and nonverbal modes in operation. Two millennia ago, ekphrasis was an exercise given to orators to make them aware that they did not deal solely with the listener's reception of words: they also dealt with a range of mental responses to bodily sensations. For writers today – possibly *less* likely than ancient orators to understand their rhetorical effect upon readers' minds – the now expanded idea of ekphrasis provides a portal into how our writing is received, and also how we make it.

What we retrieve in writing creatively are the knowledge-laden phenomena that flash onto our mind-screen, are captured by our mind-mic, get transformed at our mind tasting-table, find description in our feel-test and smell-calibration mind laboratories, and get transformed into word-bites. Brilliantly, we coax and push this information into the writing of fiction, poetry, plays, memoir, essays, etc. We relay to others what our mind processes generously provide to us.

3 Thinking and Writing, According to Writers

The writer's view of the thinking done in the writing process differs from that of the literary critic and the cognitive scientist, not just because the writer is present in their own mind to experience it happening but also because the evaluating language used and the cultural context it refers to are so unlike those of the humanities and the soft and hard sciences. This may sound self-evident but 'experts' other than creative writers (viz., philosophers, literary critics, linguists, cognitive psychologists, neuro-biologists and so forth) have been listened to more closely than writers themselves in the judgement and analysis of the brain-work required by writing. There are many arguments to challenge this state of affairs. For example, critics analysing the mind-activity revealed in literary texts prob-ably report more about their own *reading* minds, than about the writer's mind. Scientists investigating the writer's mind may not have asked the experimental questions that really matter to the writer. Some few of the scientists quoted in this book are, in fact, recognised novelists, poets and playwrights, which for me makes them particularly worth listening to (e.g. Jane Piirto, Charles Fernyhough and James C. Kaufman) while others (e.g. Todd Lubart and Scott Barry Kaufman) trained in creative arts areas (visual arts and music, respectively) and show flair and empathy in their engagement with creative writing sufficient to gain the full admiration of a creative writer (Henshon, 2016; Kaufman, 2021). An affinity with the creative arts is clearly an advantage in a scientist who studies creative arts thinking processes.

So, what have creative writers themselves revealed by their investiga-tions of thinking? I will start this chapter with an account of their view of the writing mind as a whole. They have seen the mind as a place of work, and typically have employed metaphor to examine its dynamism, its intricacy and what it feels like to labour there. I have already mentioned Rousseau's marvellous use of 18th-century stagehands' work in great Italian theatres and 'the air of disorder [that] prevails... while the scene is being changed' (Rousseau, 2008 [1782]: 111) to describe the turmoil in his mind when writing. In the mid-19th century, Edgar Allan Poe used the metaphor similarly, although he modernised it somewhat by referring to fly-tower machinery and other tricks of the theatre trade. In 'The philosophy of composition' (1846), Poe provided a sustained and detailed, thought-by-thought account of the mind-work which went into the writing of his poem, 'The Raven' (1842). He introduced the project thus:

> I have often thought how interesting a magazine paper might be, written by any author who would – that is to say, who could – detail, step by step, the processes by which any one of his compositions attained its ultimate point of completion. Why such a paper has never been given to the world, I am much at a loss to say – but, perhaps, the autorial vanity has had more to do with the omission than any one other cause. Most writers – poets in especial – prefer having it understood that they compose by a species of fine frenzy – an ecstatic intuition – and would positively shudder at letting the public take a peep behind the scenes, at the elaborate and vacillating crudities of thought – at the true purposes seized only at the last moment – at the innumerable glimpses of idea that arrived not at the maturity of full view – at the fully matured fancies discarded in despair as unmanageable – at the cautious selections and rejections – at the painful erasures and interpolations – in a word, at the wheels and pinions – the tackle for scene-shifting – the step-ladders and demon-traps – the cock's feathers, the red paint and the black patches, which, in ninety-nine cases out of the hundred, constitute the properties of the literary *histrio*.
>
> I am aware, on the other hand, that the case is by no means common, in which an author is at all in condition to retrace the steps by which his conclusions have been attained. In general, suggestions, having arisen pell-mell, are pursued and forgotten in a similar manner. (Poe, 2001 [1846]: 743)

Poe then launched into his account with a paragraph beginning 'The initial consideration was...' and continued the tract with paragraphs commencing 'My next thought...', 'These points being settled, I next bethought me...', and so on (Poe, 2001: 743–750). Rousseau and Poe share the perception that the creative mind is a messy work-place, and also a site of impressive transformations. In spite of the worrying chaos apparent in the procedures, the final production can be exquisitely seamless. Rousseau and Poe, in their centuries-old descriptions, were surprisingly close to the situation that neuroscience confirms today: the creative process does not occur neatly in a select part of the brain, or even on just one side of it. It involves a mass of complex interactions across the whole brain (Kaufman, 2020), quite legitimately characterised as a swarm of movement across a darkened stage, or a flurry of activity behind the scenes of the creative production.

In 1818, John Keats had described his mind as a different sort of work-place: a rural one. He contemplated the

...fears that I may cease to be
Before my pen has gleaned my teeming brain,
Before high-pilèd books, in charactery,
Hold like rich garners the full ripened grain;...

(Keats, 2020 [1848])

Keats said powerfully and simply that his brain was a field from which he gathered a harvest of ideas, images and emotions for his writing, and his books were the granaries (garners) where the harvest was stored. One might envy the fact that Keats' mind was such a teeming, fertile source of material and inspiration; but for a young poet apprehensive of an early death, his productive brain was a problem rather than a comfort. This sonnet, 'When I Have Fears' (written in 1818, published in 1848), not only describes a writer worrying that he might die before producing all the works he wanted to, it also traces step by step the thinking done, and the visual and emotional imagery each step conjured, in the argument being made. Keats evoked the pastoral scene of agricultural growth, the busy moments of reaping, thresh-ing, cleaning and hauling to the granary, and the swelling glory of the final product. The movement between idyll and industry accurately portrays the writer's mind involved in periods of reflection, slow growth, gathering in of perceptions, and then busy execution of the work for publication.

Henry James, in a 1911 letter to fellow writer H.G. Wells, referred to 'the great stewpot or crucible of the imagination, of the observant and recording and interpreting mind' (James, 1972 [1911]: 77). Henry's image of the 'great stewpot or crucible' seems directly influenced by brother William's 'seething caldron of ideas' (James, H., 1972 [1908]). Its associations are variously: the domestic work-place of the kitchen; the magical and somewhat threatening work-place of the witch's lair; the mythical laboratory of the alchemist; and the massive production lines of Victorian factories. Each is a place where transformation happens – raw materials become products of a higher order. Rousseau, Poe, Keats, and then the James brothers, all agree that the writer's mind is a busy site of innovation and development.

Writers' Representation of Verbal Thinking

Writers of the 16th and 17th centuries were keen observers of the mind. Elizabethan dramatists developed the soliloquy as a way to inves-tigate human psychology and utilise its power in the theatre. Shakespeare,

Kyd and Marlowe wrote famous soliloquies – among them, Hamlet's 'To be, or not to be' (c.1600) remains today the most celebrated representation of human thinking in literature:

> To be, or not to be, that is the question:
> Whether 'tis nobler in the mind to suffer
> The slings and arrows of outrageous fortune
> Or to take arms against a sea of troubles,
> And by opposing end them.
>
> (Shakespeare, 1963 [1600–1601]: III, i, 56–60)

In *Hamlet*, Act III, Scene 1, Shakespeare decided that the real stage for the play's key moment was his central character's mind. Hamlet's consideration of suicide was the crucial uncertainty on which the play's message hinged. The mind-battle Hamlet describes, involving army-like 'slings and arrows' and 'arms', is not seen by the audience but is meant by ekphrasis to be staged in their minds. We know from the Chorus's opening speech in *Henry V* (written around the same time as *Hamlet*) the power of the audience mind that Shakespeare relied upon for successful dramatic writing and stage production. Here, with Hamlet, he wanted the character's imagined battle to become the audience's imagined battle. The *actual* stage set for this scene is nondescript, a room in the castle where a variety of characters move in and out, often not seeing or hearing each other, with the performance area a shifting multi-space representing the fact that reality often presents itself to us as indistinct, plural and ambiguous. The entire play is about what goes on problematically in people's minds – how they are deceived into thinking at odds with reality: for example, how ghosts appear and forests move; how a representation of reality (e.g. a performance by strolling players) disturbs the guilty and the innocent; and how your thinking creates your destiny (which is the case for all the characters in the play). The Elizabethans saw that *thinking* was the heart of drama: it was drama itself for a character to speak their mind:

> To die, to sleep –
> To sleep – perchance to dream: ay, there's the rub,
> For in that sleep of death what dreams may come...
>
> (Shakespeare, 1963 [1600–1601]: III, i, 64–66)

The issues Shakespeare investigated in the cognitive focus of his play concerned taking steps into fearful unknowns (death, murder, suicide, engagement in battle, love). Such moments of crisis provide triggers for the writer to shift the narrative from the besieging real world to the embattled mind. Hamlet's stream of consciousness, as revealed, is an argument between voices in the mind, a source-point of ultimate conflict, the very essence of drama.

In cognitive psychology terms, this is precisely the kind of mental dialogue that neuropsychologist Charles Fernyhough (2016) describes in *The Voices Within*. While all good drama is in essence about conflict, and while Shakespeare's play *Hamlet* is essentially about conflicts in the mind, Hamlet's soliloquy describes the essential activity that occurs mentally: the cognitive vocalisation in contemplation of the individual's own being. Shakespeare had no choice really in deciding where to stage this crucial human drama about personal ethics and survival. It had to be in the mind, with his character wrangling the plural voices that typically assail one in the most challenging moral circumstances. Leon Katz says:

> The soliloquy's private conference with the self is usually working towards the resolution of a dilemma, or working through the likely obstacles to a plan of action, or – and these are the most interesting ones – are governed by battles of conscience whose fixed ideals and unspoken beliefs over-whelm practical intent. (Katz, 2002: xvi)

Speaking about the actor's challenges in delivering mind-activity to the audience, Katz says:

> Whether rational or deranged, the mind is moving from one notion to the next in a chain of changing assertions that link. Figuring the exact con-tinuity of those links can be easy – in narratives or in expositions which merely detail a sequence of ordered events. But when the connections are either muffled or random, the work of the actor begins. (Katz, 2002: xvii)

But 'the work of the actor' has already been the work of the writer. The blindness, the agony, the epiphany, the capitulation, the derangement – whatever the actor must portray – the writer has already created, analysed and staged it in their own mind.

Between William James publishing his prototype 'On some omissions of introspective psychology' (1884) and his masterwork *The Principles of Psychology* (1890), brother Henry published his novella 'The Aspern Papers' (1888). Its first-person narrative was focused on the richly detailed thinking of the anonymous narrator. The following excerpt is typical. It involves a conversation about renting an apartment in Venice. But the narrator is obsessively focused on his own mind-shifts:

> [The landlady's statement] had all the air of being a formula of dismissal, as if her next words would be that I might take myself off now that she

had had the amusement of looking on the face of such a monster of indiscretion. Therefore I was all the more surprised when she added, with her soft, venerable quaver, 'You may have as many rooms as you like – if you will pay a good deal of money'.

I hesitated but for a single instant, long enough to ask myself what she meant in particular by this condition. First it struck me that she must have had really a large sum in her mind; then I reasoned quickly that her idea of a large sum would probably not correspond to my own. My deliberation, I think, was not so visible as to diminish the promptitude with which I replied: 'I will pay with pleasure and of course in advance...' (James, 1963 [1888]: 170)

Henry James captured with great economy the rhythm of individual thinking that his brother William investigated, producing a perfectly nuanced account of 'a single instant' of mind-work: a flash of intuitive, fast-paced convergent thought that operates in response to the communication moment, to the statement, tone of voice and gestures of the interlocutor, and to the perceived nature of the situation. Henry James' ability to tease thinking apart became his trademark as a writer. He did not subscribe to his brother's notion of the divergent thought-train where carriages of different proportions might link very different objects and cause seemingly illogical jumps in thinking, but he did tease out the logicality of thinking where the fluidity of 'stream of thought' was paramount.

Henry James' novels are considered precursors to the stream-of-consciousness writers of the 20th century, some of whom, as he did, narrated in past tense and third person, had characters remembering their stream of consciousness episodes, and did not use the less-edited immediacy of present-tense mind-work as Joyce and others did. The acknowledged first novel to employ immediate stream of consciousness technique (*monologue intérieur*) was Edouard Dujardin's *Les Lauriers sont coupés*, published in 1887, translated as *We'll to the Woods No More* in 1938 and as *The Bays are Sere* in 1991. Henry James scholar Leon Edel called it 'the minor work which inaugurated a major movement' (Edel, 1990 [1957]: xxvii) principally because in 1922 James Joyce revealed to Stuart Gilbert that he had read Dujardin's short novel in 1902 and praised the book highly (Edel, 1990 [1957]: ix). Gilbert wrote:

'In *Les Lauriers sont coupés*,' Joyce told me, 'the reader finds himself, from the very first line, posted within the mind of the protagonist, and it is the continuous unfolding of his thoughts which, replacing normal objective narration, depicts to us his acts and experiences. I advise you to read *Les Lauriers sont coupés*.' (Gilbert, 2015 [1930])

Joyce's advice was excellent, of course. Dujardin's novel was considered 'trivial' and a 'stunt' (Edel, 1990: vii, x) but is better described as a daring, ground-breaking investigation of verbal and visual cognition. Here is a taste:

> ...A street; Rue de Marengo; the Louvre Stores; a dense file of carriages. Chavainne:
> — I'll have to leave you at the Palais-Royal.
> What a bore he is! Always giving one the slip like that! We are under the arcade now; walking past the shop windows; in the crowd. Better walk on the road. No, too many carriages. Bit of a crush here, but it can't be helped. A woman in front; tall, slim heavily scented; shapely figure she has, flashing red hair; wonder what her face is like; handsome, probably. Chavainne is speaking.
> — Come with me tonight to the theatre. After we can go for a stroll somewhere. (Dujardin, 1990 [1887]: 13)

Throughout the novel Dujardin orchestrates setting, action, dialogue, psychology and detailed perception in a manner suggesting he learnt the technique from masters. But in 1887 this style of writing had never been published before. While the story is about one evening in the love life of a superficial, insecure young Parisian dandy, Daniel Prince, it is also a penetrating account of his mind-activity.

Daniel's insecurities lead him into constant debate with himself and change his mind about what he should do next – an excellent example of Fernyhough's (2016: 100) 'dialogic thinking'. At the same time, a wonderfully poetic quality pervades Daniel's powers of observation. In a beautifully rendered flaneur passage (Dujardin, 1990: 84–89), his thinking combines description of the cityscape with the effects of a tune played on a hurdy-gurdy interwoven with memories from his country childhood and his insistent notions about the nature of love. It is a complex performance of a mind at work. Leon Edel, while admiring Dujardin's experiment, also took him to task for 'struggling' with the problem of descriptive detail, suggesting the author 'breathlessly moves an excessive quantity of furniture into the consciousness of his personage' (Edel, 1990: xx–xxi). For example, here is Daniel's mind-voice upon returning home:

> Here's my door; I open it; darkness; matches in their place; I strike one; careful now... the sitting-room door; I enter; the mantelpiece, candlestick there; I light it... My hat and coat on a chair; I enter the bedroom; the two double candlesticks with their storks' necks branching up; light them; it's done. My room; on the left there my bed, white in its bamboo frame;

the panel of old tapestry hanging above it; a red pattern, quiet-toned, subdued, with vague blue-violet patches on a shadowy background of deep reds and blues, all faded tints; I really must get a new mat for the dressing-room... (Dujardin, 1990: 49–51)

By complaining about the way this type of passage is set up, Edel missed the point that Daniel does *not* describe his surroundings solely as a means by which the author announces the setting for the action. Daniel's observations are coded with his judgements about himself and his Parisian society. Before readers enter his apartment with him, they have in previous pages experienced his fantasy about the wealth he wished he had and the fine things he might impress others with to maintain his aspirant social standing. Dujardin's subtlety in presenting the ironies of his character's psychology is nuanced and perceptive, both in the details the character observes and the thinking they trigger in him. But also, the technique investigates the type of narrative artifice a creative writer is compelled to employ when representing *visual* cognition. Without resort to actual visual images, seen things must be transposed into the words of verbal cognition. One can read Molly's monologue in *Ulysses* and have difficulty seeing where the scene actually happens – except, of course, in her head. Another notable difference between Dujardin and Joyce is that Dujardin indicated his cross-cuts at the end of completed sentences while Joyce did away with punctuation almost entirely. By this, they both achieve the Jamesian stream of consciousness as 'flow' as well as 'chopped up bits' (James, W., 2019: 239). Joyce did it more dramatically, perhaps – and that might be why Dujardin is less celebrated.

Dorothy Richardson is recognised as another pioneer of stream-of-consciousness narrative. In 1915, a quarter-century after Dujardin and seven years before *Ulysses*, she published the novel, *Pointed Roofs*. The earliest stream of consciousness section of her novel is Section 4 of Chapter II, where the central character Miriam first arrives in Europe (on her way to a position in Germany to teach English) and her mind-voice is reported:

It was a fool's errand.... To undertake to go to the German school and teach... to be going there... with nothing to give. The moment would come when there would be a class sitting round a table waiting for her to speak. She imagined one of the rooms at the old school, full of scornful girls.... How was English taught? How did you begin? English grammar... in German? Her heart beat in her throat. She had never thought of that... the rules of English grammar? Parsing and analysis... Anglo-Saxon prefixes and suffixes... gerundial infinitive.... It was too late to look anything up. Perhaps there would be a class to-morrow.... (Richardson, 1921 [1915]: 26–27; ellipses in original)

Richardson took authorial tracing of thoughts further than Henry James did, but not as far as Dujardin or Joyce. This was not interior monologue, because the first person 'I' was not in use; but a mind at work was implied and examined in third person. Also, it was not the thinking brain recorded in real time, as in Dujardin and Joyce, and as Hamlet's soliloquy was. It is better described as *free indirect speech*. But, just as Shakespeare did with Hamlet, Richardson turned to stream of consciousness to examine a mind in a state of anxiety, contemplating how to move forward into a dreaded new world. Admittedly, Miriam's apprehension about teaching English to German pupils is not of the order of Hamlet's consideration of suicide but, for both characters, contemplation of steps into fearful unknowns provided the trigger for the writer to shift the narrative focus to the character's mind.

For comparison, Gertrude Stein's novella, 'Tender Buttons' (1912), was published three years earlier than *Pointed Roofs*. The backstory here was that William James' student Stein, among all the American writers of the early 20th century, had a powerful influence not only on writing but on other art forms, too. In her highly experimental work, she performed more radical versions of the representation of thinking than even her mentor William James elaborated. She was the real successor to his explorations because she investigated extreme cases of the thinking flow – the dissimilarity of components in the stream of consciousness. In 'Tender Buttons', she was not interested in the logicality of convergent thinking: she was committed to the divergent and the 'fringe'; she tested the links and affinities between thoughts, the possibilities of fragmentation, the variable connections between the carriages in the James-train. For example, in 'Tender Buttons' she described two rooms in a house – a young woman's bathroom and bedroom – and what goes on in them. But also, she described 'rooms' in the mind:

> The time when there is not *the question* is only seen when there is a shower. Any little thing is water.
> There was a whole collection made. A damp cloth, an oyster, a single mirror, a mannikin, a student, a silent star, a single spark, a little move-ment and the bed is made. This shows the disorder, it does, it shows more likeness than anything else, it shows *the single mind that directs an apple*. All the coats have different shape, that does not mean that they differ in colour, it means a union between use and exercise and a horse. (Stein, 2003 [1912]: 292; my italics)

'The question' was that of the relatedness of subjective perception to the nature of the world. When one is in the shower, water is all one sees. One's world is water, a veil of water (thus, 'Any little thing is water'). Can I be certain about my stating this? Of course not. But I did have that connected thought, and it is exactly what Stein wanted me to do – have a divergently linked thought created in the periphery of my thinking train. Also, when she said, 'the single mind that directs an apple', maybe she wanted me to think of Eve, or perhaps think about particular notions of orderliness that women might have, or particular kinds of thinking women might have, as opposed to those that males have, or perhaps she wanted me to remember that horses like apples. And on the colour and shape of coats related to horses, I see a woman dressed for dressage, perfectly ordered. That's definitely order personified – a dressage woman! But who knows? The author of the thoughts in the text wanted to provoke thoughts in the reader, yet not author them. Additionally, Stein used a mind-voice that focused on suggestion of visual thinking, yet also other senses: touch (having a shower), taste (eating an apple), perhaps even smell (one's clothes after exercising a horse).

Stein's achievement in 'Tender Buttons' was to record the divergent connections her mind provided her with, and to find a narrative form by which they might be transferred to the reader. She hoped to contact the reader's own thinking at deeper levels than those produced by conventional narrative text. Most writing, Stein provocatively suggested, is fashioned by authors for the *limitations* of readers' thinking, so that they do not have to do too much mental work. Stein threw the burden of meaning-making back to the reader. *Here's how my mind is working while writing*, she said. *What do you make of it yourself, dear reader?*

After Hamlet's soliloquy, the most famous monologue in literature is the last chapter of James Joyce's *Ulysses* (1922), Molly's soliloquy:

> ...Ive a mind to tell him every scrap and make him do it in front of me serve him right its all his own fault if I am an adulteress as the thing in the gallery said O much about it if thats all the harm ever we did in this vale of tears God knows its not much doesnt everybody only they hide it I suppose thats what a woman is supposed to be there for or He wouldnt have made us the way He did so attractive to men then if he wants to kiss my bottom Ill drag open my drawers and bulge it right out in his face as large as life he can stick his tongue 7 miles up my hole as hes there my brown part then I'll tell him I want £1 or perhaps 30/- Ill tell him I want to buy underclothes then if he gives me that well he wont be too bad I dont want to soak it all out of him like other women do I could often have written out a fine cheque for myself and write his name on it for a couple of pounds a few times he forgot to lock it up besides he wont spend it Ill let him do it off on me behind provided he doesnt smear all my good drawers O I suppose that cant be helped Ill do the indifferent 1 or 2 questions Ill know by the answers when hes like that he cant keep a thing back I know every turn in him Ill tighten my bottom well and let out a few smutty words smellrump or lick my shit or the first mad thing comes into my head... (Joyce, 1966 [1922]: 929–930)

Joyce took the male author's ultimate dangerous journey: into the female mind. He delivered a woman character's thinking beyond where Stein went sexually, but still very much in keeping with what William James proposed. Joyce fashioned his version of the James-stream by removing the linguistic elements (punctuation, spelling, accepted sentence structure) to produce a more intimate flow of language seemingly akin to the mind-voice we all continuously think in. He did not go as far as Stein with James-train associative leaps. Molly thinks in conventional ways (even though her behaviour may not have been conventional for her time): her thoughts are linked quite rationally and sequentially. But James went deeper into consciousness than literary writing went before, by examining sexual thinking, an area previously reserved for pornography. In doing this, Joyce transformed writing about thinking and introduced its particular 20th-century cast – brutally honest confession, no-holds-barred transmission of details, scandalous (for the time) public revelation of what really occurs in private minds, and confirmation that Freud's sexy unconscious might be more conscious than everyone had chosen to acknowledge (see, for example, Horgan, 2012). When a writer deals with the mind honestly, it relies on the fact that there is no society-imposed moral, legal or ethical stricture. Stuff goes on in minds perennially that would close down news channels were they to broadcast it and that would lead to prosecutions. If we were tried in court on our thinking, we might all end up in gaol. Mostly, when released at all, the stuff of private minds is kept confidential in therapists' notes, or it finds its way into creative writing.

Virginia Woolf saw the new century's reading needs in her essay 'Modern fiction', published in 1919. In contrast to what the mass of writers were doing in contrived, conventional, 'like-to-life' novels of the time, Woolf sought verisimilitude not of the external world, but of the inner:

> Look within and life, it seems, is very far from being 'like this' [i.e. what it is customarily seen to be]. Examine for a moment an ordinary mind on an ordinary day… the accent falls differently from of old; the moment of importance came not here but there; so that, if a writer were a free man [sic] and not a slave, if he could write what he chose, not what he must, if he could base his work upon his own feeling and not upon convention, there would be no plot, no comedy, no tragedy, no love interest or catastrophe in the accepted style, and perhaps not a single button sewn on as the Bond Street tailors would have it. Life is not a series of gig lamps symmetrically arranged… Is it not the task of the novelist to convey this varying, this unknown and uncircumscribed spirit, whatever aberration or complexity it may display, with as little mixture of the alien and external as possible? We are not pleading merely for courage and sincerity;

we are suggesting that the proper stuff of fiction is a little other than custom would have us believe it. (Woolf, 2009 [1919]: 9)

Experimental writers for the rest of the 20th century followed Woolf's lead. Even the conventional genres reacted. Working in the crime genre while also striving for literary excellence and innovation, Elmore Leonard made his characters' criminal and sexual thinking accessible to readers. This quote is from his novel, *Gold Coast* (1980):

> Maguire thinking of a snowbanked Durant Mall in Aspen, deep powder on the high slopes, the rich ladies in their snow-bunny outfits. Then thinking of the Pier House in Key West, sitting out on the deck with a white rum and lemon, six in the evening. Places out of the past. Thinking of fifteen hundred bucks and what they could scrounge out of the lockers, maybe two, three hundred more each. Thinking of islands and palm trees... (Leonard, 1982 [1980]: 27)

Leonard tracked the jumps in thinking that we all do – from experience to fantasy, from past to present to future, from one place to another. He presented character mind-moves not dissimilar to those that Stein and Joyce followed, in that they are based in personal perception, emotion and divergent reflection – but also, he inserted a cogency driven by crime genre requirements. He appealed to audience mind participation, and he reminded us as creative writers that the drama in the mind is forever a personal performance to be monitored. Leonard worked with the James-stream to show that pace of thinking engages the reader, and that flow of ideas keeps the crime novel going. But the key thing about the fragments of thinking linked together here is this: they are *visual*. You might not notice at first how significantly visual they are, because our ideas about stream of consciousness tend toward conceiving of narrative as verbal. But narrative is visual, too. We can tell stories with a picture or a series of pictures. Hamlet's thoughts were partially visual (the 'slings and arrows', the 'sea of troubles') as were Molly's ('kiss my bottom Ill drag open my drawers and bulge it right out'), but Leonard's Maguire thinks exclusively in images and they, too, build narrative and comprise story.

Finally, it is useful to look at a 20th-century experimental writer putting the mind-voice onto the page while at the same time using the mind-eye as assistant: Raymond Federman.

Raymond Federman's *Double or Nothing* (1972), one of many 20th-century experimental works aimed at portraying the inner voice, used verbal thinking in collaboration with visual thinking to create an

Figure 3.1 The narrator considers how to write the story (Federman, 1998 [1972]: 70)

intimate sense of mind-activity. The page shown in Figure 3.1, in spite of having expansive white spaces, suggests the confined nature of the 'closed box' in which the internal dialogue takes place. The narrator here is the fictional author of the work, and his interrogative mind-conversation replicates a typical moment in any creative writer's thinking as they choose among details in developing a work. Notably, this fictional author considers history, cultural and political ramifications, family significances, personal survival, reader confusion, the overall structure of the work at hand and, ultimately (it turns out), the voice of the narrative itself – all in the momentary verbal thinking of writer choosing the name of a character.

Writers' Representation of Visual Thinking

In the 'Stream of thought' chapter of *Principles of Psychology* (1890), William James discussed visual thinking and used a diagram to capture his argument (see Figure 3.2). He interpreted the figure in the following way, in words:

> Let *A* be some experience from which a number of thinkers start. Let *Z* be the practical conclusion rationally inferrible from it. One gets to the conclusion by one line, another by another; one follows a course of English, another of German, verbal imagery. With one, visual images predominate; with another, tactile. Some trains are tinged with emotions, others not; some are very abridged, synthetic and rapid, others, hesitating and broken into many steps. But when the penultimate terms of all the trains, however differing *inter se*, finally shoot into the same conclusion, we say and rightly say, that all the thinkers have had substantially the same thought. It would probably astound each of them beyond measure to be let into his neighbor's mind and to find how different the scenery there was from that in his own. (James, W., 2019: 269)

The flow of thinking in the James-train, William James said, can comprise dissimilar 'carriages', or elements – they can be an ongoing

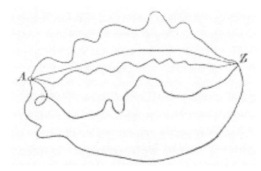

Figure 3.2 William James' 1890 representation of his argument about visual thinking (James, W., 2019: 269)

combination of verbal bits, visual bits, tactile or emotional bits, joined together. The way in which two individuals think towards the same outcome seldom means that the two have thought in exactly the same way. How we each think reflects the typical practices of our mind, the way habitual thinking pathways have been constructed by the bodies we have, the influence of our environments and cultures, and our personal experience. Science shows that some thinkers use mental images more than others; some use words in the head more; some even have a preference for tactile or emotional cognition. For example, when seeking to describe a ball, the tactile thinker might shift to the feel of a ball in the hand, while the emotional thinker might re-experience how they felt when they were hit on the head by a ball. Of interest to creative writers is whether there might be some particular *way* to move thinking forward in their own mind – some best-practice version of the James-train, a more effective combination of elements in the shifting flow of thought – which might make one a more engaging communicator, a writer better able to grasp and hold the attention of the reader and the praise of the critic. Early ideas about the best-practice version of writing-thinking were studied by ancient rhetoricians in ekphrasis (creating images in the audience mind) and prosopopoeia (creating voices in the audience mind).

Ekphrasis was developed by the Greeks and Romans to teach the skills of *description* – or how to get a listener/reader to see in their head what the orator/author was seeing in theirs (Krauth & Bowman, 2017). The first text quoted in the study of ekphrasis is always Homer's *Iliad* (c.1200 BCE), the section where the god Hephaestus forges in his smithy an astounding blacksmith artwork – a shield for Achilles to use in the battle of Troy. The description of the shield has led to discussion over millennia mainly because it is not just a description of a beaten metal object and an example of notional ekphrasis but is also an allegorical narrative representing the sum total of Greek understanding of the nature of the civilised world in Homer's time. Homer fashioned the worded version of Achilles' shield to represent war and peace, violent and harmonious human behaviour, the work of cities and rural areas, the progress of the seasons, and a celebration of the natural life-cycle. In total, Homer's description of the shield comprised a performance of universal visionary proportions. I cannot think of anything to equal it other than Whitman's extraordinary vision of humanity in sections of 'Song of Myself' (*Leaves of Grass*, 1855).

There are various translations of *The Iliad*, but I will use Alexander Pope's 1715 work. Pope was a significant writer who brought to the task not just translation skills but also writerly insight. Pope began the section in Homer's Book 18 thus:

Then first he formed the immense and solid shield;
Rich various artifice emblazed the field;
Its utmost verge a threefold circle bound;
A silver chain suspends the massy round;
Five ample plates the broad expanse compose,
And godlike labours on the surface rose.
There shone the image of the master-mind:
There earth, there heaven, there ocean he designed;
The unwearied sun, the moon completely round;
The starry lights that heaven's high convex crowned...

(Homer, 1909: 344–345; my italics)

Pope's translation acknowledged Hephaestus' powerful passion in his creation. Inspired by Homer's description of the creative process, Pope injected into his translation a writerly perception about what goes on in the mind of the creative worker, insisting that 'There shone the image of the master-mind' – a conception of the process unrecognised by more recent translators, e.g. Fagles (Homer, 1998: 483). Pope revealed his writer's understanding of ekphrastic mind-work – not ekphrasis seen from the translator's or critic's viewpoint but from the creator's viewpoint. Pope returned to it 62 lines later:

Now here, now there, the carcases they tore:
Fate stalked amidst them, grim with human gore.
And the whole war came out, and *met the eye*;
And each bold figure seemed to live, or die.

(Homer, 1909: 346; my italics)

Pope was fully aware that the shield text, whether in beaten metal or in words, must 'me[e]t the eye'. That included not just the eye of those in situ but also the mind's eye of the reader, and, prior to that, the mind's eye of the writer. The mind's eye is not just a readerly concept – it is part of the creative process:

Next this, *the eye the art of Vulcan leads*
Deep through fair forests, and a length of meads;
And stalls, and folds, and scattered cots between;
And fleecy flocks, that whiten all the scene...
A figured dance succeeds... a comely band
Of youths and maidens, bounding hand in hand.

(Homer, 1909: 347; my italics)

Again, Pope stressed that the shield-maker's work led the observer's eye forward to seeing the setting, the action and the characters. Homer and Pope were both keen to explain the importance of visual thinking to the writing and reading process in this earliest major ekphrastic exercise in our literature.

Homer's vision of the world on a shield – his most influential study of the mind's visual workings – focused on the dramatic conflict of battle (and also on the harmony of dance, battle's opposite). Battle is a major human experience, but usually a vicarious one. Although most people are not ever in the front line, it is fearful nevertheless in the imagination. Shakespeare's most incisive investigation of human visual thinking also involved battle. In the Prologue to *Henry V* (1599–1600) the Chorus explains how the visual imagination works for both writer and audience/reader:

O for a Muse of fire, that would ascend
The brightest heaven of invention,
A kingdom for a stage, princes to act
And monarchs to behold the swelling scene!
Then should the warlike Harry, like himself,
Assume the port of Mars; and at his heels,
Leash'd in like hounds, should famine, sword and fire
Crouch for employment. But pardon, and gentles all,
The flat unraised spirits that have dared
On this unworthy scaffold to bring forth
So great an object: can this cockpit hold
The vasty fields of France? or may we cram
Within this wooden O the very casques
That did affright the air at Agincourt?
O, pardon! since a crooked figure may
Attest in little place a million;
And *let us*, ciphers to this great accompt,
On your imaginary forces work.
Suppose within the girdle of these walls
Are now confined two mighty monarchies,
Whose high upreared and abutting fronts
The perilous narrow ocean parts asunder:
Piece out our imperfections with your thoughts;
Into a thousand parts divide one man,
And make imaginary puissance;
Think when we talk of horses, that you see them
Printing their proud hoofs i' the receiving earth;
For 'tis your thoughts that now must deck our kings,
Carry them here and there; jumping o'er times,
Turning the accomplishment of many years

Into an hour-glass: for the which supply,
Admit me Chorus to this history;
Who prologue-like your humble patience pray,
Gently to hear, kindly to judge, our play.
(Shakespeare, 2021 [1599–1600]: I, Prologue, 1–34; my italics)

This is Shakespeare's brilliant analysis of how theatre relies on audience visual mind-work. '[L]et us… On your imaginary forces work', he said. 'Piece out our imperfections with your thoughts… And make imaginary puissance,' he said. 'Think when we talk of horses, that you see them', he said. For it is the audience's visual thinking that must carry the play. With their physical eyes they see ten actors, but with their mind's eye they see ten thousand soldiers. With their physical eyes they see a stage but, with their mind's eye, a battlefield, and a kingdom. And this is the case for poets and fiction writers, too. As creative writers we rely on the 'imaginary forces' that the visual-thinking reader supplies to expand and animate the worlds, characters and action that we describe in mere, yet powerful, words.

In the article on ekphrasis that Chris Bowman and I wrote (2017), Chris pointed out that contemporary writers 'have given their own metaphors to the ekphrasis process… talking about it as "telepathy", as a "portal", a means to transport the reader into the subject's presence':

Stephen King in *On Writing* (2000) describes telepathy as 'what writing is', and follows with a concise description of a table with a red tablecloth. On the table is a cage, and in the cage is a white rabbit munching a carrot-stub. The numeral 8 is marked on the rabbit's back in blue ink.

It's an eight. This is what we're looking at, and we all see it… we *are* together. We're close.
We're having a meeting of the minds.
I sent you a table with a red cloth on it, a cage, a rabbit, and the number eight in blue ink. You got them all, especially that blue eight. We've engaged in an act of telepathy. (King, 2000: 78–9)

This is King's way of explaining the ekphrastic act between the writer and the reader – the transfer of a mental image from one mind into another, collapsing time and distance to bring visual life to the words chosen by the writer. (Krauth & Bowman, 2017)

Creative writing's use of visual thinking is not confined to isolated exercises in ekphrasis: it is an integral part of writing itself. Among the creative writers who confirm the importance of visual thinking in their writing process, a small number in the last 50 years have analysed it extensively (including Arthur Koestler, Joan Didion, Italo Calvino and Sue Woolfe). The small number may be due to the fact that the creation of visual imagery in the mind is so inherent to writing that it is taken for granted and goes largely unnoticed.

The relationship between the processes of visual thinking (imagining, daydreaming, fantasising, remembering) and our mind's creation of visual imagery from external sources (our normal mode of seeing) is not perfectly understood by science, although current experimentation finds the former to be a 'less precise' version of the latter (Breedlove *et al.*, 2020: 2,211). For writers when they write, the visualised scene or object is often a remembered one – and we are aware that we regularly get details *wrong* (a word I use advisedly), which may not be a problem for certain kinds of creative writing in the popular mind but is a serious problem for creative writers themselves. In my own case, I want to get it right: that crook in the elbow as the character moved; that particular adjacency of building to hillside; that specific odour in the air that night. When I imagine an *action scene* for transposing into fiction, I see it on my mind-screen (my 'mental cinema', as Calvino [2016: 102] called it), I run it like a movie (I am 'the cinema operator who works the projection machine', as Koestler [1989 (1964): 180] said), and I transfer it via my fingertips to sentences on the page. When I describe a character's *physical appearance* or a *setting* – subjects with more static features – I visualise a portrait of the character, or a landscape or cityscape, on my mind-screen and then 'paint' it in words. However, when I wish to reproduce the visual thinking of a character, i.e. the image a character would voice externally via the voice speaking inside their head (as Joyce did with Molly Bloom), then I utilise my own verbal- and visual-thinking modes together, I see in my head the image the character sees and I write it as if the character's mind dictates it for transcription. Taking the viewpoint of any character involves me seeing the seeing of outer and inner worlds that my character is doing on my authorial mind-screen, as if through the mind's eye of the character.

In his epic study of creative thinking in the arts and sciences, *The Act of Creation* (1964), novelist Arthur Koestler quoted the biographies and autobiographies of scientists such as Einstein, Faraday and Kekulé, and creative practitioners such as S.T. Coleridge, to demonstrate that visual thinking performed a major role in invention, new discoveries and innovative thought processes. Doing his analysis in the 1960s, Koestler was keen to relate creative writing to scientific thinking at the time (which allowed a more significant role for the unconscious than we might tolerate today). He suggested that, in most 'truly original acts of discovery... the "seeing" is in fact imagining; it is done in the mind's [eye], and mostly the unconscious mind's eye' (Koestler, 1989: 200). Then, he said, 'the unconscious mind's eye', which does the groundwork, hands over to the *conscious* mind's eye in order to continue the process of shaping the original insight into a viable, communicable work:

> The visualizer [e.g. the writer]... feels his [sic] way around a problem and strokes it with his [mind's] eye... trying to fit it into some convincing or elegant shape; he plays around with his vague forms like the couturier with his fabrics, draping and undraping them on the model. (Koestler, 1989: 182–183)

As the inventor/author develops the innovative idea and builds the work, their visualising processes resemble daydreaming, or the conscious running of film clips in the cinema in the head: they are 'the cinema operator who works the projection machine, and [are] the audience at the same time' (Koestler, 1989: 180). Different from the sleeping dreamer, who is 'the spectator passively watching the sequence of images', the active daydreamer (the inventor with a project to develop, the creative writer with a work to progress) is in charge of the visualising process (Koestler, 1989: 180). Visualisation is dominant in the developmental phase, Koestler thought, because words are not fluid enough for truly creative thinking – they have cultural and linguistic baggage attached:

> Words... crystallize thought; they give articulation and precision to vague images and hazy intuitions. [But] the creative act... presupposes a relaxing of the controls and a regression to modes of ideation which are indifferent to the rules of verbal logic, unperturbed by contradiction, untouched by the dogmas and taboos of so-called common sense. (Koestler, 1989: 173, 178)

Koestler captured the value of divergent thinking and Jamesian 'fringe' logic here. Mind-image affiliations can take thinking further into creative possibilities, enhancing creative and scientific outcomes alike.

There is a long list of writers who talk about an initiating *visual* image at the beginning of a writing project – John Dryden, S.T. Coleridge, Mary Shelley, D.H. Lawrence, Jean Cocteau, Tennessee Williams, Stephen Spender, Joan Didion and so on. Henry James (1908) recalled a dinner party conversation where a guest told a story and an image for a novel flashed into life and 'glimmered' in his mind:

> There had been but ten words, yet I had recognised in them, as in a flash, all the possibilities of the little drama of [my novel *The Spoils of Poynton*], which glimmered then and there into life... (James, H., 1972 [1908]: 73)

John Hawkes (1965) described the spark for his novels thus:

...in each case I began with something immediately and intensely visual – a room, a few figures, an object, something prompted by the initial idea and then literally seen, like the visual images that come to us just before sleep. (Enck & Hawkes, 1965: 148)

E.L. Doctorow (1986) answered the question, What comes first?, in relation to his novel *Loon Lake* with:

I had these opening images of a private railroad train on a single track at night going up through the Adirondacks with a bunch of gangsters on board, and a beautiful girl standing, naked, holding a white dress up in front of a mirror to see if she should put it on... I kept thinking about the images. (Doctorow, 1986)

Beyond initial motivating visualisations, fiction writers confirm that they develop, manipulate and narrativise their starting images using visual and verbal transformations to get written text onto the page.

In an 'Author's introduction' (1831), Mary Shelley gave an account of the initiating moments for her novel *Frankenstein* (1816):

Many and long were the conversations between Lord Byron and [Percy] Shelley, to which I was a devout but nearly silent listener. During one of these... [t]hey talked of the experiments of Dr. Darwin... who preserved a piece of vermicelli in a glass case, till by some extraordinary means it began to move with voluntary motion... [and it made me think] perhaps the component parts of a creature might be manufactured, brought together, and endued with vital warmth. (Shelley, 2013 [1831])

Shelley went on to describe how, later during the night, further visualisations developed her thinking about the themes and plot for the work:

Night waned upon this talk, and even the witching hour had gone by, before we retired to rest. When I placed my head on my pillow, I did not sleep, nor could I be said to think. My imagination, unbidden, possessed and guided me, gifting the successive images that arose in my mind with a vividness far beyond the usual bounds of reverie. I saw – with shut eyes, but acute mental vision, – I saw the pale student of unhallowed arts kneeling beside the thing he had put together. I saw the hideous phantasm of a man stretched out, and then, on the working of some powerful engine, show signs of life, and stir with an uneasy, half vital motion. (Shelley, 2013 [1831])

As the images proliferated to suggest successive scenes and plot action, they also gave rise to possibilities for character psychology and emotion:

> Frightful must it be; for supremely frightful would be the effect of any human endeavour to mock the stupendous mechanism of the Creator of the world. His success would terrify the artist; he would rush away from his odious handywork, horror-stricken. He would hope that, left to itself, the slight spark of life which he had communicated would fade; that this thing, which had received such imperfect animation, would subside into dead matter; and he might sleep in the belief that the silence of the grave would quench for ever the transient existence of the hideous corpse which he had looked upon as the cradle of life. He sleeps; but he is awakened; he opens his eyes; behold the horrid thing stands at his bedside, opening his curtains, and looking on him with yellow, watery, but speculative eyes. (Shelley, 2013 [1831])

Immersed in her mind's imagery, Shelley was more than inspired, she was horrified – and the fictional juggernaut she had created impinged on her actual surroundings:

> I opened [my eyes] in terror. The idea so possessed my mind, that a thrill of fear ran through me, and I wished to exchange the ghastly image of my fancy for the realities around. I see them still; the very room, the dark parquet, the closed shutters, with the moonlight struggling through, and the sense I had that the glassy lake and white high Alps were beyond. I could not so easily get rid of my hideous phantom; still it haunted me. I must try to think of something else... (Shelley, 2013 [1831])

Thankfully for readers over the next two centuries, Shelley did not reject the succession of mental images that produced her horror masterpiece:

> Swift as light and as cheering was the idea that broke in upon me. 'I have found it! What terrified me will terrify others; and I need only describe the spectre which had haunted my midnight pillow.' On the morrow I announced that I had thought of a story. (Shelley, 2013 [1831])

She then tells us: 'I began that day with the words, "It was on a dreary night of November..."'. In this case, the author made from mental imagery a strongly developed sequence of scenes that are ready for transcription into a narrative draft. The process is ekphrastic in that she turns visual thinking in her mind into verbal thinking produced as words on the page.

Mary Shelley's forensic account of the provenance for her literary classic was very different from husband Percy's grandiose explication of the poet's mind in his 'A defence of poetry' (written 10 years earlier) with which she was undoubtedly familiar. Percy Shelley talked about his writing as 'indeed something divine':

> A man cannot say, 'I will compose poetry'. The greatest poet even cannot say it: for the mind in creation is as a fading coal which some invisible influence, like an inconstant wind, awakens to transitory brightness... (Shelley, 2001 [1840]: 713)

The contrast between Mary's modest report on her (feminine) brain-work and Percy's pontification about the (man-) poet's psyche evokes many lines of discussion, but a peaceful way to deal with it is this: among that group of writers who discussed the possibilities of the Horror genre in a house beside Lake Geneva in 1816, it was not the luminaries Percy Shelley or Lord Byron who responded by writing the first horror classics but the non-celebrities: Mary Shelley and John Polidori (who published the first vampire story, *The Vampyre*, in 1819). Perhaps this indicates something about why so little has been produced by writers observing their own thinking – what Poe (2001 [1846]: 743) called 'the autorial vanity': Mary was willing to get down to the nitty-gritty of describing activity in her feminist head because she saw it as part of the work about women's writing that needed to be done; and Dr Polidori, as a medico inclined to think about the body generally in relation to the mind (his medical thesis was on oneirodynia – nightmares and sleep-walking) could not but see horror writing's focus on mind and body. While the celeb boys, Percy and Byron, saw themselves superhumanly inspired – heirs to literary thinking about muse-driven afflatus – Mary and John looked at a real issue: what went on in their thinking.

Joan Didion, in her article 'Why I write', published in *The New York Times* in 1976, focused on 'the act of writing' and said: 'I write entirely to find out what I'm thinking, what I'm looking at, what I see and what it means' (Didion, 1976: np). Specifically, what she looked at were images in her mind. The idea of writing to find out what it is one is thinking is perhaps unsurprising for the author writing from verbalised thoughts, but it takes on a different dimension when the writer is thinking visually; it strongly suggests an ekphrastic process where the pictorial element provided by the mind is being transcribed into words:

> *What is going on in these pictures in my mind?*
>   When I talk about pictures in my mind I am talking, quite specifi-cally, about images that shimmer around the edges... You can't think too

much about these pictures that shimmer. You just lie low and let them develop. You stay quiet. You don't talk to many people and you keep your nervous system from shorting out and you try to locate... the shimmer, the grammar in the picture. (Didion, 1976; italics in original)

Didion treats her mind-images with care since they are precious resources for her writing. Almost as if they were works of art themselves, she applies a 'Do Not Touch' policy and waits to feel their power growing, as one might do in a gallery confronted by a masterpiece to get maximum value out of the experience. Significantly, she seeks to find in these pictures their word equivalences:

Just as I meant 'shimmer' literally I mean 'grammar' literally. Grammar is a piano I play by ear... All I know about grammar is its infinite power. To shift the structure of a sentence alters the meaning of that sentence, as definitely and inflexibly as the position of a camera alters the meaning of the object photographed. Many people know about camera angles now, but not so many know about sentences. The arrangement of the words matters, and the arrangement you want can be found in the picture in your mind. The picture dictates the arrangement. The picture dictates whether this will be a sentence with or without clauses, a sentence that ends hard or a dying-fall sentence, long or short, active or passive. The picture tells you how to arrange the words and the arrangement of the words tells you, or tells me, what's going on in the picture. *Nota bene.*

It tells you.
You don't tell it. (Didion, 1976; italics in original)

Didion's insistence that the angles the writer takes in viewing the mind-image dictate the sentence structures, and that the visual-thought arrangements dictate the word arrangement, reflect Koestler's 'couturier with his fabrics'. Of course, writers are individuals and one person's process is not necessarily shared by another. But here, it seems, Didion captured 'the act of writing' in creative forms exactly. How you see it (scene, action, character) in your mind, and how you treat it, drives how you say it in your writing. But this is not an exact science, it is a very individual one: the degree to which you direct the action in the mind-film, or allow it to direct you, is a fascinating part of the uncertainty of writing.

Another incisive contributor to the analysis of visualisation in the creative writing process was Italo Calvino. In 1984, in the lectures he prepared for delivery at Harvard University, he described his process of conceiving and developing a story:

> For me, then, the first step in the conception of a story occurs when an image that comes to mind seems, for whatever reason, charged with meaning, even if I can't explain that meaning in logical or analytical terms. As soon as the image has become clear enough in my mind, I begin developing it into a story, or rather the images themselves give rise to their own implicit potentialities, to the story they carry within them. Around each image others come into being, creating a field of analogies, symmetries, juxtapositions. At this point my shaping of this material – which is no longer purely visual but conceptual as well – begins to be guided by my intent to give order and sense to the story's progression. Or, to put it another way, what I'm doing is trying to establish which meanings are compatible with the general design I want to give the story and which are not, always allowing for a certain range of possible alternatives. At the same time, the writing, the verbal rendering, assumes ever-greater importance. I would say that as soon as I begin to put black on white, the written word begins to take over, first in an attempt to match the visual image, then as a cohesive development of the initial stylistic impulse, until little by little, it rules the whole field. Then it is the writing that must guide the story towards its most felicitous verbal expression, and all that's left for the visual imagination is to keep up. (Calvino, 2016: 109–110)

Calvino's lecture on 'Visibility' in *Six Memos for the Next Millennium* (1988) is a *tour de force* in analysis of the ekphrastic aspects of the creative writing process, and a ringing endorsement of James-train processes. In a richly analytical work, Calvino's summation is contained in this statement:

> Let's say that various elements come together to form the visual part of the literary imagination: direct observation of the real world; phantasmal and oneiric transfiguration; the figurative world as transmitted by the culture at its various levels; and a process of abstraction, condensation, and interiorization of sensory experience, which is as crucial for visualizing as it is for verbalizing thought. (Calvino, 2016: 116)

Analysing Calvino's 'Visibility' lecture in the light of recent developments in neurological science and brain-scanning, Skov *et al.* (2007) say:

> ...neither visual imagery, nor the ability to represent visual images in a linguistic form, can be said to be uniquely creative phenomena; both mechanisms are commonly employed in everyday communication. What sets some types of visual imagery and linguistic representations apart as

'imaginative' or 'creative' must be a qualitative difference: some verbal images are more 'meaningful', 'difficult', or 'exciting', than others, and some linguistic representations are more 'successful' or 'poignant' than others. We may therefore speculate that both visual imagery and linguistic formulation are rooted in general neurocognitive mechanisms, and that some external 'executive' mechanism (or mechanisms) is capable of influencing these processes. (Skov *et al.*, 2007: 192)

For teaching and learning creative writing skills, this is a useful statement. It suggests that the way see normally and translate those mind-images into everyday words, are among the key skills in the writing process. But the value-added factor the writer provides, over and beyond normal seeing, is the '"executive" mechanism' that manipulates the image on the mind-screen.

Writers' Representation of Other Sense Modes

For writing sensory detail, we receive information in the mind through sense channels and process it into words. While creative writers rely predominantly on visualisation and verbalisation to create character, depict action, describe setting, and generally get the drama of a work across, they also know that conveying the full range of sensory experience – including touch, taste and smell – plays an important role in fiction, poetry and scripts.

The convincing depiction of the senses powerfully draws reader and audience in. 'When specific senses are employed,' says Jessica Page Morrell, 'the reader infers meaning' because 'details anchor the reader in a place, a moment, or a scene, and specificity deepens the reader's understanding of a scene' (Morrell, 2006: 179). The same can be said for the writer's understanding of what *they* are doing. Writerly engagement with sensual data in the brain leads to: greater immersion in the character's experience (both psychological and emotional); better understanding of the effects and consequences of action; enhanced evocation of memory, emotions and social encounters; and richer interaction with environments in the setting. 'In writing that depends on verisimilitude,' novelist John Gardner (1991 [1983]: 22) has said, the writer 'argues the reader into acceptance [by use of] precision of detail':

This kind of documentation, moment by moment authenticating detail, is the mainstay not only of realistic fiction but of all fiction... The realist [writer] must authenticate continually, bombarding the reader with proofs [while the non-realist writer uses] authenticating detail more sparingly, to give vividness to the tale's key moments... [We as reader] not only respond to imaginary things – sights, sounds, smells – as though they were real, we respond to fictional problems as though they were real: we sympathize, think, and judge. (Gardner, 1991: 23, 25)

Gardner situates authentic rendering of sensory experience at the heart of fiction's effectiveness. However, in my reading of his classic *The Art of Fiction: Notes on Craft for Young Writers*, he stopped short of mapping what the writer must do consciously to effect the conversion from sense input to authentic words to stimulate the vivid exchange between writers' and readers' minds.

The requirement that convincing sensory detail be presented involves inviting challenges for the creative writer in the task of describing what their senses-fuelled mind presents them with in what might be termed 'other sense images'. How *well* can I describe this perfume/stink? This delicious/sickening taste? This alluring/repulsive texture? How *fully* can I connect with the reader by diving into the descriptive detail I fashion with words? Neuroscientific studies acknowledge, for example, that, in the case of a person's sense of smell, compared with other animals the '[r]educed [human] repertoire of genes for olfactory receptors is compensated by the great capacity of human brain processing' (Sarafoleanu *et al.*, 2009: 197). That greater capacity supplies the power of words to assist the smelling process:

Language and speech... play an important role in the perception and discrimination of the odors. Although, we find it hard to describe an olfactory sensation, this is very important in defining and bringing the smell in the cognitive part of consciousness. (Sarafoleanu *et al.*, 2009: 197)

While our 'lower order' senses – smell, taste and touch – rely significantly on our powers of mind-description and the transfer of the incoming sensation to descriptive inner visuals and language, there has been in the writing field little consideration of the process of ekphrasis beyond its application *from visual into verbal*. Other creative arts – for example, music, film and photography (Krauth & Bowman, 2017: 14–15) – talk about ekphrastic translations between, say, music and photography or film and haptics. As Bowman and I argue, this broadened idea of ekphrasis – between any two or more of the senses – is closer to the ancient rhetoric concept than is the narrow literary version that refers exclusively to a painting becoming a poem. Neuroscience has looked at 'ekphrastic' mind-manoeuvres in the form of synaesthesia, but writers themselves have mostly ignored analysing it, with Vladimir Nabokov the notable exception (Bouchet, 2020: 255ff).

Novelist Janet Burroway said to writers on behalf of readers:

A detail is 'definite' and 'concrete' when it appeals to the senses... If you write in abstractions or judgments, you are writing an essay, whereas if you let us [the readers] use our senses and form our own interpretations, we will be involved as participants in a real way... A detail [of description] is concrete if it appeals to one of the five senses; it is significant if it also conveys an idea or judgment or both. (Burroway, 2011 [2003]: 22, 23–24)

Nailing the concrete detail is central to creative writing. Burroway points to the fact that narrative details can be conveyed both intellectually and artistically: via writing oriented towards the verbal-focused performance that happens in our heads, or via writing that embraces further our full range of sensual thinking. Concrete detail in creative writing instruction is often examined under the popular heading of 'Show Don't Tell', but exactly what that axiom involves remains debatable. Choosing between what goes on in our minds as 'telling', and what happens there as 'showing', is rarely teased apart. I suggest in terms of this book's thesis that 'telling' involves writerly mind-activity that purports to anticipate the reader's mind-work and fashions the text with more convergent James-train linkages, whereas 'showing' includes the writer trusting more to their personal affiliative linkages and allowing the reader greater participation by making their own.

Writers use seeing and hearing most often in the normal practice of writing, but there are times when creative writers must give authentic taste, touch and smell experiences to characters and at the same time satisfy the expectations readers have in those situations. Regarding taste, the following three examples demonstrate various authorial mind-processing strategies, each of which turns the taste of an orange into a poem. Nate Flying Owl's 'Ode to an Orange' (2013) provided a response that didn't rise above conventional breakfast table thinking and perhaps the normal saying of grace: 'With another slice in my mouth/I savour it with gratitude/And I pray to God, thanking Him/For the taste of the sublime orange' (Flying Owl, 2013). Brian F. Kirkham's 'Ode to an Orange' (2014) goes further, with the writer anthropomorphising the fruit and placing it in a larger social context, but still he does not mind-hack the taste of the orange to elevate it above conventional description: 'each colourful segment you hold/bursts in the mouth/quenching the driest thirst... you are served to the guests/In a Cafe where the visitors/are truly impressed' (Kirkham, 2014). To enjoy a master of mind-work in action we can experience Pablo Neruda's 'Ode to the Orange' (1954): 'when

I penetrate/your contours, your waters,/praise/your women,/and see how your forests/sway... I understand that you are,/planet,/an orange,/a fruit of fire' (Neruda, 1996 [1954]: 105).

Neruda takes a course similar to Kirkham, by socialising the experience of taste, but he extends it much further. The abundant experience of tasting the orange becomes the abundance of experiencing the world, including particularly Neruda's political heartland of Chile from which he was exiled. The succession of passionate visual images in a memory-rich, highly personalised description evokes emotions and excitement for the writer, as too for the reader. In this manufactured synaesthetic process (see below) the unexpected depiction as 'a fruit of fire' becomes, in the complex maze of associations, a new insight into taste. Probably no reader of this poem ever had the exact same response to eating an orange, but Neruda made the taste universally significant through a powerful physicality and intellectuality, all of which required concentrated mind-activity that related taste sensations to affiliated sensual, emotional and political ideas and changed conventional responses to highly personalised ones. Neruda's poem shows how a hard-won writerly metaphor can become a potent readerly experience.

Like Neruda, D.H. Lawrence produced a series of fruit poems, the most famous of which is his erotic 'Figs' (1923): 'The fig is a very secretive fruit./As you see it standing growing, you feel at once it is symbolic...' (Lawrence, 1966 [1923]: 42). Lawrence evokes for me the fruit's sensual taste and texture because I have eaten figs and the transference of taste associations from his mind to mine works well, especially because he evokes James-train associations rather than literal, uninspired taste sensations. On the other hand, I have never eaten a medlar or a sorb-apple, so in reading his 'Medlars and Sorb-Apples' (1923) I cannot rely on my memory in the same way: 'I love you, rotten,/Delicious rottenness./I love to suck you out from your skins.../Autumnal excrementa;/What is it that reminds us of white gods?' (Lawrence, 1966: 42). On the basis of his advertisement, some readers won't bother with tasting these Lawrentian fruits, but I am hooked and want to try. I am reminded of one of the few products sold in the world today that depends upon taste description for its sales – wine. Wine-tasting reviewers are required to concentrate their minds on the task of putting taste into words. And while we may be amused by ekphrastic attempts such as 'Hints of banana accompanied by hessian highlights' these creative writers of the wine industry are faced with the same task facing literary creative writers.

Similarly, few modern adults have had the tactile experience of running and rolling naked on a wet English hillside, but Lawrence seduced us with a range of touch experiences in *Women in Love* (1921) to convince us that we can participate:

> He went through the long grass to a clump of young fir-trees, that were no higher than a man. The soft sharp boughs beat upon him, as he moved in keen pangs against them, threw little cold showers of drops on his belly, and beat his loins with their clusters of soft-sharp needles. There was a thistle which pricked him vividly, but not too much, because all his movements were too discriminate and soft. To lie down and roll in the sticky, cool young hyacinths, to lie on one's belly and cover one's back with handfuls of fine wet grass, soft as a breath, soft and more delicate and more beautiful than the touch of any woman; and then to sting one's thigh against the living dark bristles of the fir-boughs; and then to feel the light whip of the hazel on one's shoulders, stinging, and then to clasp the silvery birch-trunk against one's breast, its smoothness, its hardness, its vital knots and ridges – this was good, this was all very good, very satisfying. (Lawrence, 1967 [1921]: 119)

As with many writers who attempt to depict touch, Lawrence relied often on visually-oriented description of the action of touching ('soft sharp boughs beat upon him', 'feel the light whip of the hazel on one's shoulders', 'clasp the silvery birch-trunk against one's breast') but also he used affiliative, mode-crossing, non-cliché description such as 'pricked him vividly' and 'fine wet grass, soft as a breath' to conjure touch with more complexity in the reader's mind. These descriptions cause the reader to become aware of a different sort of ekphrasis that their mind is going through: 'Vivid? Isn't that about seeing? Breath? Is that about touch?' W.E. Williams said Lawrence's writing was 'detonated rather than composed', and that in his later life, Lawrence 'was suspicious and impatient of revision, a process he thought liable to tamper with the spontaneity' of his writing (Williams, 1966: 9). The keys to Lawrence's writing were intimacy and immediacy.

Lawrence considered that the connections the mind made in its fringe-thought activities were not achieved by laboured deduction and editing: they were the outcome of intense, concentrated raids on the mind's processing of incoming sensory material in real-time. Lawrence wrote: 'I have always tried to get an emotion out *in its own course*, without altering it. It needs the finest instinct imaginable, much finer than the skill of craftsmen' (Lawrence, quoted in Sagar, 2007: 25). Lawrence eschewed overworking the sites of conscious activity where a passionate sensual experience (which he linked to 'emotion') was raised and captured. He famously did not trust what his mind presented him with; he trusted instead his body: 'The mind can assert anything and pretend it has proved it. My beliefs I test on my body, on my intuitional consciousness, and when I get a response there, then I accept' (Lawrence, 2004 [1927]: 208). Lawrence's suspicion of the mind was an outcome of how mind was conceived in the early 20th century, when the Cartesian mind–body split was not yet defeated. Were he alive today, Lawrence would be heartened by the fact that science and creative writing caught up with his thinking to acknowledge that the mind is not easily distinguishable from the body: his 'intuitional consciousness' is today recognisable as embodied cognition, a legitimate recorder of experience.

In Patrick Süskind's novel, *Perfume* (1985), the olfactory sense is the absolute focus of the work. The protagonist Grenouille is born amid 'a putrefying vapour, a blend of rotting melon and the fetid odour of burnt animal horn' and gutted fish which 'stank so vilely that the smell masked the odour of corpses' (Süskind, 2006 [1985]: 5). His mother was insensible to these stink invasions and the boy grew up with a nasal capacity so astonishing that 'everyday language would soon prove inadequate for designating all the olfactory notions that he had accumulated within himself':

> ...why should smoke possess only the name 'smoke', when from minute to minute, second to second, the amalgam of hundreds of odours mixed iridescently into ever new and changing unities as the smoke rose from the fire... or why should earth, landscape, air – each filled at every step and every breath with yet another odour and thus animated with another identity – still be designated by just those three coarse words. All these grotesque incongruities between the richness of the world perceivable by smell and the poverty of language were enough for the lad Grenouille to doubt that language made any sense at all... (Süskind, 2006: 29–30)

Grenouille's problem is Süskind's problem, too: how does the writer convey myriad subtle variations for a sense so rarely described in detail? Another novelist, Peter Ackroyd, said of Süskind's work:

> That this is in every sense an olfactory novel gives a striking sensory immediacy to the fiction itself. *Perfume* is a historical novel but one in which the sheer physicality of its theme lends it an honorary present tense. (Ackroyd, 1998 [1986])

By working with sense details, the writer engages with readerly mind-activity at the moment of reading. Süskind's book analyses many ekphrastic strategies – for example, a scent coming from behind a wall has a visual effect on the protagonist: 'To be sure, it... came from a redheaded girl, there was no doubt of that. In his olfactory imagination, Grenouille saw this girl as if in a picture...' (Süskind, 2006: 197). An ekphrasis from scent to touch (with visual, touch and taste highlights) can also occur:

> That night as he lay in his cabin, he conjured up the memory of the scent... and immersed himself in it, caressed it, and let it caress him... and he made love to it and to himself through it for an intoxicatingly deliciously long time. (Süskind, 2006: 220–221)

The author's mind-hack of his own – and his readers' – ekphrastic processes worked so well that in the end, when the brilliant perfumier-cum-serial killer is dismembered by a scent-crazed crowd, we are sorry to see him go: he has told us so much about our understanding of the most primal of our senses and our sensual thinking.

The above examples show that in order to record sensory experience, a typical writerly tactic is to resort to metaphor. Metaphors usefully access concepts in complex description situations – 'All the world's a stage' (Shakespeare), 'Books are the mirrors of the soul' (Woolf), 'All religions, arts and sciences are branches of the same tree' (Einstein). Descriptions of taste, touch or smell can be difficult for writers since the English language has less developed vocabularies for these senses compared with the vocabularies for seeing and hearing. We can make the less differentiated senses more vivid by likening them to more often experienced or more easily referenced encounters: a bright taste, a dazzling touch, a booming odour. This process involves literary pseudo-synaesthesia, which is a controlled metaphor-making version of the inter-sense transfer (synaesthesia) that neuroscientists detect in some 3% of the population who are synesthetes, automatically hearing colours, tasting sounds, touching smells, etc., because of their brain-wiring (Baron-Cohen & Harrison, 1997: 8–11; Ramachandran & Hubbard, 2001). When, in 1590, John Donne wrote that 'a loud perfume' betrayed the presence of a surreptitious lover, in 'Elegy 4 [The Perfume]' (Donne, 1896: line 41), or Keats in 1818 suggested the reader 'taste the music of that vision pale' in the poem *Isabella* (Keats, 1884: line 392), or Baudelaire mentioned 'odours... mellow as oboes' in 'Correspondences' (Baudelaire, 1993 [1857]), or when Vladimir Nabokov talked about his 'coloured hearing' in *Speak, Memory* (Nabokov, 1989 [1951]: 34), the reader must deal with the writer's capture of what happens in their willing, or unwilling, cross-mode mind process. These marvellous metaphors exercise the reader's cross-sense connector capability in a process explained by William James as an adaption of the 'fringes and haloes of relations' (James, W., 2019: 264) that surround a normal thought. With any thought there is 'that original halo of obscure relations… spread about its meaning', James says (2019: 275–276), but with the pseudo-synaesthetic metaphor the possibilities of this halo are particularly attractive to the creative writer and can be exploited to produce highly effective words on the page.

Stephen Ullmann, in *The Principles of Semantics* (1957), proposed a hierarchy of senses in metaphoric language based on his reading of the Romantic poets. He said (Ullmann, 1963 [1957]: 280): 'transfers tend to mount from the lower to the higher reaches of the sensorium, from the less differentiated sensations to the more differentiated ones', i.e. from touch to sight in the following order: touch > taste > smell > hearing > sight. Not all have agreed with him, a particular suggestion being that hearing is top of the hierarchy (Tsur, 2007). Psychologist Raymond W. Gibbs, who defined metaphor as 'a mental mapping' occurring between 'diverse domains of experience' (Gibbs, 2011: 113), suggested that metaphor is, in itself, a sign of creativity and 'may be part of many higher-level creative activities such as problem-solving, decision-making, and novel language use' (Gibbs, 2011: 115). According to P.L. Duffy: 'The study of synaesthesia is helping researchers to understand the variety of ways that the human brain can process and code language' (Duffy, 2020); and some neuroscientists suggest synesthetes 'might have extra wiring between adjacent brain regions', while other investigators propose 'they simply have extra activity in the existing wiring' (Weir, 2019). The take-away for creative writers is that making striking metaphor involves the task of summoning together parts of our brains that do not normally work together, which do not necessarily have go-to neural pathways established, but which have a capacity to form when called upon. For some of us, doing the mind-work that produces the drama and memorability of a coloured smell or an audible taste is demanding work; for others it comes more naturally.

The mind's relationship with what goes on around it generally, with what has recently happened specifically, with what occurred a greater time ago and is remembered, and with what it does habitually, is complex. The writer has to come to terms with and exploit – but not be exploited by – what happens in their mind in comprehensive and multidimensional ways. The nature of thinking has always been foundational to the nature of writing. Good writing in all forms is an outcome of authorial thinking pressured by the conventionality of cultural views but also triggered by input from individual drives and perceptions. While it might be said that writing in conventional forms is less taxing on the writer's mind than is writing experimentally, nevertheless writers since ancient times have explored how putting together *unconventional* narrative or verse can reveal the normal working of the mind. Narratives made from fragments of sensory experience, from multimodal combinations of physical touch, taste, smell, hearing and sight elements – all such experimental writing provides evidence of one thing above all: the human mind at work (Krauth, 2016a, 2019). It interests me that non-linear writing is classed as 'experimental' in critical literature, but that there is little acknowledgment that the divergent thinking behind it was part of William James' formative psychology, which described how we think *normally*. Creative writers' experiments may not have followed science-genre write-up rules, but they have contributed significantly to how we understand our senses operate in our brains.

4 The Mosaic Mind: Writing and Divergent Thinking

The Mosaic Mind

I once watched a wall-tiler at work. He picked shards of broken tile from a messy pile, then sized and positioned them *mosaically* as the facing for a house front. I felt I had much in common with this man. His process paralleled divergent thinking in the writer's process. I presumed he was accustomed to laying tiles in regular rows, but, in this case, he avoided positioning like-to-like pieces in linear queues. He wanted a collage, wanted unlike against unlike; but he spent thoughtful time considering the uneven adjacencies, the contrasting colour tones, the shapes of gaps between the shards. He seemed to have an overall pattern in mind but seemed also to allow it to develop in its own way from the materials at hand. I had earlier seen the architectural artwork of Antoni Gaudí's smashed and recycled crockery in Barcelona, and the effects achieved by the *pique assiette* decorators of the Wat Arun temple in Bangkok. I was inspired and amazed by the way these building designers made whole compositions from broken bits. (For comparison, George Orwell thought Gaudi's work hideous whilst Evelyn Waugh thought it genius [Waugh, 1930: 309]). And I was delighted when I read that in 1896 the Italian author, Gabriele D'Annunzio, helped a tiler lay a pavement in Venice and the experience later inspired the writing of his best work, *Notturno* (1921). D'Annunzio's novel, a collection of fragments written during his period of bed-ridden temporary blindness after WW1, was fashioned exclusively in the studio of the mind, so to speak – the product of focused cognitive work denied its normal access to visual and other sensory input (Hughes-Hallett, 2013: 17; Krauth, 2019). But it was still directed by an overall pattern which, I would say, followed the author's knowledge of how identifiable James-train associations occur.

The tiling metaphor works for how we think when combining elements in creative works. From my own *actual* tiling experience – I am a frustrated home builder – positioning a straight row of tiles requires thinking akin to that for writing a conventional essay: the elements must line up and be aesthetically pleasing in ways dictated by linear, 'rational' logic. This is convergent thinking in action. When we don't want the tiles in linear formation but still want to create a meaningful and pleasing overall effect – albeit by unconventional logic, more complicated geometry and more adventurous placement – the thinking required is divergent. Linear thinking sees this behaviour as 'whimsical' or 'irrational', but collage thinking has its own rationality and deals with complex links across the whole pattern, which may not involve slavish repetition in local links within the pattern. Again, in Jamesian psychology terms, 'fringes and haloes of relations' (James, W., 2019: 264) are recognised and allowed to take a role in the creative process. No matter how they write, writers at all times manipulate discrete elements, whether they have linearity or fragmentation in mind: writing is always a putting-together of bits – the words for a sentence, the sentences for a paragraph, the paragraphs for a chapter, and so on, within a whole product. But how we arrange those bits, those adjacent words, sentences, paragraphs, etc., is subject to the different kinds of mind-work we apply to them.

In following James-train thinking through various writers' strategies, I have so far suggested that convergent thinking – where adjacent segments of thought line up habitually and sequentially – may be best-suited to explicatory genres such as the essay, the scientific paper, and journalism. On the other hand, divergent thinking operates well for experimental modes such as radical discontinuous and fragmented narratives (Krauth, 2016a: 110–119), which treat thinking in ways not limited by scientific or essayistic ambition to formally discover an answer or come to an unambiguous conclusion. William James' ideas derived from his study of people struggling to find answers to problems and having difficulty putting bits of thinking together to make meaningful sense in their lives. Conflicted thinking occurs often in literary characters' minds across all genres, although generally it is delivered by writers in conventional ways with a logical narrative voice making a good argument about the problem at hand. Aware of this, experimental writers regularly ignore conventional literary meaning-making when describing *actual* mind-work and they use non sequiturs, confusing illogicality, and 'abrupt cross-cuts and transitions', as William James called them, to build the idea of *authentic* thinking patterns. But these fragmented depictions of the thinking mind are difficult for readers to engage with, so creative writers usually avoid the experimental approach and turn the actual divergent nature of thinking into a more palatable 'fashioned' convergence that fits the ethical idea that writing should instruct and help readers, or at least communicate with and entertain them, not thrust them further towards despair.

Creative writing explores on behalf of the readership's aspirations, so it need not be beholden to the rationality of scientists or essayists. Just the same, most writing attempts to scale back and streamline the commonly divergent thinking that goes into its making (as Poe emphasised). The normal mind is essentially a confusing mosaic of thoughts before we, as writers, edit it down to a logical or semi-logical sequence of thinking. Whether it be Henry James giving account of the manoeuvring mind of the nameless narrator in 'The Aspern Papers' (1888), James Joyce tracing the provocative thinking of Molly Bloom in *Ulysses* (1922), Virginia Woolf cogitating her way towards delivering a university lecture in *A Room of One's Own* (1929), Raymond Federman's fictional author strategising the plot possibilities for the novel he himself is part of in *Double or Nothing* (1972) or Jonathan Safran Foer tracking the traumatised thoughts of nine-year-old Oskar Schell in *Extremely Loud & Incredibly Close* (2005), writers have been obliged by publishing and reading conventions to make sense of thinking for readers. While also depicting everyday regular thinking, creative writing tends towards exploration of thinking in crisis. As writers, we seek drama and conflict in plots and character development, and for this we observe the nature of our own thinking in crises. At the best of times, in mentally stable thinking, we do not necessarily put thoughts together in cogent ways; but in crisis, the nature of thinking is even less likely to be linear. Unfortunately for scientists, they are obliged to look objectively at *other* people's thinking in trauma and must rule out their own irrationality of mind as evidence. But creative writers have the advantage of studying their own flailing, and sometimes failing, minds.

Virginia Woolf, a writer particularly keen to map the workings of her own and others' minds, said of the variety of entries in Katherine Mansfield's journal published in 1927: 'nothing could be more fragmentary; nothing more private' (Woolf, 2011 [1927]: 63). Then, having acknowledged the privately fragmentary world of the mind, she said: '[Mansfield] is a writer; a born writer. Everything she feels and hears and sees is not fragmentary and separate; it belongs together as writing' (Woolf, 2011: 63). There are many ironies evoked here centring around the notion that brilliant, normal, and sadly non-coping thinking are all fragmentary to one degree or another: writers, according to Woolf, are concerned with and professionally required to put the fragments of the mind together into meaningful mosaics (a task which sometimes brings a writer's sanity undone). Luke Thurston (2008) introduced his essay on the

Portuguese author Ferdinand Pessoa's fragmented self by quoting Virginia Woolf talking about the mosaic nature of thinking:

> 'We're splinters &c mosaics; not, as they used to hold, immaculate, monolithic, consistent wholes,' wrote Virginia Woolf in her diary [in 1924]. We should pause over the two terms neatly joined by Woolf's ampersand, for they suggest different, perhaps antithetical, ways of thinking about the self. A mosaic, after all, is precisely an arrangement of 'splinters', an assembly of fragments into a new totality or consistency. Today we might see in that difference an index of contrasting aspects of Woolf's own writing-identity: stylistic innovation on the one hand, personal fragmentation on the other. But Woolf's sense, at least in her diary note, of the difference between an outdated 'whole' self and a modern fragmentary one is not an anxious but an enthusiastic, almost jubilant one. The demise of the 'monolithic' Victorian ego was, in her eyes, something to be celebrated, for it corresponded to a liberation from the 'ill-fitting vestments' of nineteenth-century prose, with its conventional structures of plot, character, and 'plausibility' tailor-made to constrict or misrepresent reality and falsify the 'myriad impressions' of the human psyche. For Woolf this aesthetic liberation was, moreover, not simply a matter of literary style. It encompassed a whole new contact with life... (Thurston, 2008: 175–176)

Engaging with the mosaic nature of thinking brings us closer to understanding what humans, and our lives, really are. Needing to find a way to describe how a character puts their thoughts together, or how a character sees a series of events unfold, creative writing is not obliged to reproduce rational, cogent thinking. Hamlet's presentation of a rational account of his thinking in crisis – 'To be, or not to be: that is the question' – sounds like the beginning of a lecture by a learned professor – even though it is seen by other characters in the play as madness, and so, too, by critics. But Hamlet's rational presentation of thinking asks us today: How mad was he really? Why could he put forward such a beautifully cogent case about the crisis in his mind? Why wasn't he simply jabbering?

Gertrude Stein's writing illustrated the point that tracking thinking as it really happens to an individual makes serious demands on readers and audiences, and especially on creative writers themselves. In the process of metacognition, we need to be hypersensitive to our own mind-leaps and we need to remember or record them as well. Similarly, with our observation of others, we need to recall and record just which shifts, jumps and cross-cutting their speech and action reveal to us about what their minds are doing. If a writer really wants to trace character thinking in crisis as also at other times, then writing in logical, cogent or fluid-flowing prose and poetry seems to us now not entirely adequate or genuine. But when we write about character mind-activity in anything other than conventional, logic-driven terms, we as authors must engage the reader in the ways that Shakespeare outlined in the *Henry V* Prologue – we need the reader's mind to direct and stage-manage the mosaic of diverse scenes, characters, speeches, feelings, ideas and actions that we depict as happening on our character's mind-stage – in order to make the text work.

Robert Coover's (1969) modern classic, 'The Babysitter', relies on convergent and divergent thinking to relay fragments of experience and mind-activity in an account of what happened to seven people on a night in the American suburbs in the 1960s. The sequence of events as narrated is neither straightforward nor necessarily rational: while convergent thinking is required to deal with each separate fragment in the mosaic of the story, the collaged progress of the narrative challenges even the best reader's divergent thinking. Here, Coover played with a kind of writing that could in fact be published on playing cards or, with later technology, be delivered by random-choice hypertext means. He gave his instructions for reading a similar story, 'Heart Suit' (published on ersatz playing cards in 2005), as: 'The thirteen heart cards may be shuffled and read in any order…' (Coover, 2005). This challenge to conventional thinking about narrative structure applies to many experimental works over the last century (Krauth, 2016a: 110–119). Each of them invites the reader to pick up shards and arrange them in a random or preferred order. Coover demonstrated that reading requires divergent James-train thinking, where adjacency of segments is brought into focus, and that a variety of thinking occurrences goes on: writers place the elements of their probably highly divergent thinking in sequences meant to streamline into some sort of cogency for the reader; the reader picks up these pieces and re-fits them according to the skills and proclivities of their own reading mind. Regarding the narrative segmentation in Coover's 'The Babysitter', any confused reader can work out the narrative progress if they try hard enough (probably involving a lot of note-taking) to find the convergence of meaning within the divergence of presentation (and there are cheat websites available to assist). But few readers spend time on understanding mosaics produced by experimental writers. 'The Babysitter', now a story half a century old, ushered in the kind of thinking required for 21st-century hypertext and hypermedia writing, and at the same time demonstrated that James-train fragmented thinking could generate a sophisticated literary story structure.

In the complex way of writing commonly called *writing in fragments*, the writer takes into account the mind-activity of the reader as part of the authorial thinking process. The writer does not massage and conventionalise links between segments of narrative to make sure the reader gains a crystal-clear understanding of how the meaning-flow operates. The writer does not rationalise the text with a linear logic but leaves it to the reader to put bits together: to make thoughtful links, to apply their own thinking-jumps across the vacant gaps. At the same time, the writer's understanding of their own mental process influences the way the text is presented. Writers undertake studied consideration of how people think in order to publish this sort of material. For example, Jonathan Safran Foer in *Extremely Loud & Incredibly Close* perceived how the reader's mind will work to combine conventional and unconventional text, conventional and unconventional photographic images, along with highly unconventional page-use (e.g. the inclusion of a photographic flip-book) to produce a mainstream novel. Foer's insight into the mind-work by which readers interpret the links in the James-train was exemplary. He said of his methods:

> To me fiction is… about… a different mode of thinking than the conversational everyday reasonable motive. And reasonable thinking brings you to a reasonable conclusion, whereas unreasonable thinking or openness to the unconscious, whatever you want to call it, visceral thinking, can lead you to other places which to me are the terrain of art. It distinguishes art from journalism or rhetoric, just in the ways that it can be unreasonable. (Foer, quoted in Testard, 2012)

Foer struggled for terms to describe what goes on in his writing mind and he reached out to the old-school idea that the 'unconscious' drives what the writer does. But he corrected this quickly by calling it 'visceral thinking', something to do with the nervous system, with what the brain responds to in a lively and conscious manner. In itself, Foer's response in this interview was an example of James-train thinking.

Walter Benjamin described the mosaic nature of thinking in 1928:

> Tirelessly the process of thinking makes new beginnings, returning in a roundabout way to its original object. This continual pausing for breath is the mode most proper to the process of contemplation… Just as mosaics preserve their majesty despite their fragmentation into capricious particles, so philosophical contemplation is not lacking in momentum. Both are made up of the distinct and the disparate… (Benjamin, 1998 [1928]: 28)

Benjamin explained that our thinking is not fixed to a linear sequence. His description of normal thought rhythms correlates with William James' ideas about the thought stream: 'an alternation of flights and perchings'

(James, W., 2019: 243). Normal thought tends to be divergent, disordered and often random in its jump-cuts, cross-cuts and deviations. But conventional readers don't want the rhythms of normal thought in the texts they read: linearised texts (with the divergent thinking of the author's process edited out) appeal to them because in that form the putting of the fragments together has already been done, the logical connections have already been made. On the other hand, unconventional texts can usefully expose the mind-work of the writer, or their characters, but only a committed minority of readers is willing to follow along and make its own connections.

Yet why should experimental writing (and mosaic art and collaged film, etc.) be so hard to read? Psychology tells us that we think divergently, or 'experimentally', when required (Runco, 2011a: 400–403), and also that divergent thinking is part of the 'brainstorming' sort of activity that we regularly employ in critical, convergent-thinking situations (Russ & Dillon, 2011: 70–71). Creative writers often think in crisis mode because they write about characters in crisis, about the world in crisis, and about dramatic aspects of life. The disruptions of crisis engage the reader. Writers don't write the smoothness of the mundane as part of a narrative except as a counterpoint to the fragmentation of crisis. The mundane, commonly recognised by readers as boring, requires a lesser order of mind-activity both to write and to read.

The options available to creative writers in terms of how faithfully to reproduce the actual mosaic nature of thought might be summed up as follows:

- *Good experimental writing* takes depiction of thinking to places and processes not normally evoked – and hopefully makes it accessible, albeit with hard work involved, for the reader.
- *Good genre and non-fiction writing* has done the hard work about depiction of thinking for readers well before they start reading.
- *Good literary writing* runs a course between these two, balancing the real nature of thinking against a depiction of thinking that the reader can engage with without excessive discomfort.

Creative writers must choose the level of engagement at which to pitch their work, whether that involves significant critical mind-shifts for the reader or something less challenging and more comfortable. We tend to write convergently when we seek to instruct or entertain. We tend to write divergently – in more radically fragmented sequences of thought – when we critique the literary *status quo* or probe deeper into the real nature of experience. This corresponds with the idea that we might select and place the shards of our house wall tiles more in straight lines, or less so. Do we want to guide the observer to see an overall pattern to the mural? Or shall we leave them to their own thinking devices?

'Things Fall Apart': The Rise of Mosaicked Writing

The kind of thinking that writers do when they write in fragments involves the notion (surprising to western cultural eyes) that spaces between events are as substantial as the events themselves. For the west, a space is *empty* as well as *negative*: it *lacks* because it is not filled. For example, the concept of (outer) space being full of 'dark matter' confuses because we cannot imagine empty space being anything other than *empty*. We have similar difficulty with reading experimental works that refuse to explain in a pedestrian manner the gaps in their narrative sequences. Literary scholar, Wolfgang Iser, and others have argued about how readers fill the gaps in fragmentary narratives (see Krauth, 2019): their debate involved contentious western ideas that had at base not only the question of how we think when we read but also how we think at all – the same territory that William James covered in the late 19th century. How do we associate segments of thinking in a sequence separated by spaces, or in a sequence comprised of non sequiturs? How do we make links between bits? One of the difficulties in answering this question is that western culture does not have the concept of, nor any word for, *space with substance.*

In eastern cultures, however, the concept exists that space does have substance, and it does particular sorts of work. Andrea Day explains that the Japanese word *ma* translates as *space* or *interval* but also describes:

> a consciousness… the simultaneous awareness of form and non-form deriving from an intensification of vision. (Day, 1998)

Day wrote this in the context of architecture, but the idea exists in other Japanese art forms, including painting and poetry. Haiku poet, Alan Summers, said: '*Ma* is a fascinating area to explore, and along with white space/negative space, there is not enough written about these subjects…' (Summers, 2012). He quoted Hasegawa Kai:

> *Ma* is at work in various areas of life and culture in Japan. Without doubt, Japanese culture is a culture of *ma*. This is the case with haiku as well. The 'cutting' (*kire*) of haiku is there to create *ma*, and that *ma* is more eloquent than words. That is because even though a superior haiku may appear to be simply describing a 'thing', the working of *ma* conveys feeling (*kokoro*).

In contrast to this, Western culture does not recognize this thing called *ma*. In the literary arts, everything must be expressed by words. But Japanese literature, especially haiku, is different. As with the blank spaces in a painting or the silent parts of a musical composition, it is what is not put into words that is important. (Kai, quoted in Summers, 2012)

Further confirmation of this concept, where a space can be 'more eloquent than words', comes from 6th-century BCE Chinese writer, Laozi, who said: 'Thirty spokes unite at the single hub;/It is the empty space which makes the wheel useful' (Laozi, 2019: 109). To recognise that the spaces in the spoked wheel contribute as much to the function of the wheel as do the spokes themselves, is not part of western thinking.

Fragmented writing genres are as fundamental to the development of literatures in western cultures as they have been to those in the east. In the west, texts comprising sequences of fragments date back to Hippocrates' 412-fragment *Aphorisms* (400 BCE), which introduced systemised medical practice as we know it via a treatise where the reader was left to put the significant 'puzzle' pieces together (Hippocrates, 1994–2009). Stoic philosopher Epictetus' influential *Enchiridion* (135 ACE), written in 51 fragments, employed the aphoristic genre to discuss values and ethics in early western civilisation (Epictetus, 1948). In the east, Laozi's 81-fragment *Tao Te Ching* (2006–2019 [6th century BCE]) fundamentally influenced the development of Chinese philosophy and religion, and Sei Shōnagon's series of private fragments, *The Pillow Book* (2006 [c. 1002]) – an 'apparently crazy quilt of vignettes and opinions and anecdotes', as Meredith McKinney described it – is now seen as an eastern classic due to its exuberant documentary engagement with slices of life in the Japanese Heian court circa 1000 ACE (McKinney, 2006: ix, xv, xvi). A study of writing in fragments in expository, memoir, biography and fiction genres from classical, medieval, 20th-century and recent authors – with a focus on the aphorism, the feuilleton and hypertext fragmented forms – indicates that writers fashion a fragmented text so as to hand over a significant part of its meaning-making to the reader. In doing so, the writer manipulates the work as a mosaic of words, images and other sense impressions and writes 'directions for meaningful reading' into the spaces between the fragments (Krauth, 2019).

In the 19th century, the English Romantic poets and the French Symbolists began in earnest the kind of experimentation that would free up language to give a more authentic account of the perceptive individual's experience in the world. Poems presented as fragments by Coleridge, Wordsworth and Shelley, for example, suggested not only the increasing complexity and brokenness of the world but also portrayed the greater intensity of engagement that focus on a fragment of experience could engender. In finding that old conventions did not adequately communicate new ideas about the world in the early 20th century, creative writers sought narrative approaches that made more intellectual sense of the uncertainty of the times for contemporary audiences. One key strategy, with potential to reflect the newly perplexing nature of experience after WW1, was to fully invest in divergent thinking and its manifestation in highly fragmented narrative. Reasons for this came from life itself: individuals had to piece together understanding of life by interpreting and amalgamating sequences of often incomprehensible and mainly non-sequitur occurrences that life afforded, and they needed to decipher/parse/construe the gaps between those randomly presented events for clues to building a personal picture. Meaningfulness never arrived in sequitur sequences. The linearity of old-fashioned thinking (and the texts it produced) was a literary, political and religious subterfuge, it seemed; a form of oppression; a hoax. No lone individual was given access to an overview of life, so was vulnerable to ready-made versions of the meaningfulness of existence prepared cleverly by others who usually operated on behalf of religions and ideologies. With the idea that God was dead, and thus no cogent editing of thinking according to Christianity was available, 20th-century consciousness became more aware of divergent thinking. Without God, authentic meaning-making became a *personal* and *individual* putting together of conflicting ideas, clashing standards, and incongruous perceptions. People were, more than previously, forced to edit their own thinking, not have it pre-edited for them.

The Dadaists and the Surrealists, for example, worked on the project of finding a better way to represent the newly disintegrating world where politics, religion and morals seemed to be crumbling. In 1918 Tristan Tzara publicised the Dada cut-up writing method, where a narrative's constituent parts were separated and rearranged randomly to form a new narrative that may have had cogent, convergent thinking at its base originally but now required maximum divergence in thinking to be read. In 1920 Andre Breton and Philippe Soupault published *The Magnetic Fields* (Breton & Soupault, 1997 [1920]), an automatic writing experiment which announced that the Surrealists pursued the idea of writing in a manner *specifically* to avoid the logics of conventional linear narrative in order to overthrow the learnt rephrasing applied by 'rational' convergent thought. Following these, in 1922 James Joyce published the seemingly chaotic novel *Ulysses* in which the anti-hero, Leopold Bloom, represented the poignancies, complexities and weaknesses of humanity in the new century precisely because it so faithfully investigated the actual way people did their thinking. W.B. Yeats summed things up in his poem 'The Second Coming' (written in 1919 and first published in 1920): 'Things fall apart; the centre cannot hold / Mere anarchy is loosed upon the world...' (Yeats, 1989 [1920]). Yeats saw how the technological atrocities of WW1 had exploded conventional certainty about how the world worked and, along with it, conventional ideas about faith, cognition and writing.

The writer who best theorised writing in mosaic mode in this period was Walter Benjamin. His collections of insightful fragments, *One-Way Street* (1916 [1928]) and *The Arcades Project* (2002, published posthumously), did not ostensibly apply convergent thinking to reordering the bits of his note-like perceptions into conventional narrative. Clearly, *The Arcades Project* reads like a suitcase full of scraps, but *One-Way Street* is a crafted mosaic of 59 fragments covering approximately 75 pages where the pieces comprise a set of commentaries on the sights taken in and the thoughts experienced by a walker in a modern city. They have titles such as: 'Filling Station', 'Toys', 'Hardware' and 'Stand-Up Beer Hall', but they are far from being simple descriptions of a cityscape:

> In *One-Way Street* the [writer's] mind is constantly at play, thinking, dreaming, free-associating, not distinguishing between the trivial and the world-historical, the modern mind trying to walk and psychoanalyse itself at the same time... Benjamin is tapping into an essential modern impulse, to remake the world out of attractive, invaluable fragments. (Marcus, 2016: xxv)

While Benjamin theorised that the fragmented narrative replicated 'the process of contemplation' (Benjamin, 1998: 28), Michael W. Jennings finds an additional reason for the author's attentiveness to the form – the feuilleton writing by which Benjamin augmented his often desperately low income. The feuilleton was a section in European newspapers and magazines designed to entertain the reader: 'a short literary composition often having a familiar tone and reminiscent content... The word is a diminutive of the French *feuillet*, meaning "sheet of paper," and ultimately derives from Latin *folium*, meaning "leaf".' (Merriam-Webster, 2019).

> Many of the pieces in *One-Way Street* first appeared in the feuilleton section – not a separate section, but rather an area at the bottom of every page – of newspapers and magazines, and the spatial restrictions of the feuilleton played a decisive role in the shaping of the prose form on which the book is based. (Jennings, 2016: 2)

The feuilleton was the written fragment inserted under a line which indicated that a news article had finished. It was a space-filler, a footnote to events; it occupied a space in the paper where a reading between the lines of the rest of the news could occur. While the newspaper proper was written and organised according to convergent thinking, it was nevertheless an amalgam of ideas controlled by the editor. For inclusion of feuilleton pieces, the editor could think as randomly as he liked:

> Where else but... 'under the line,' the pronounced bold stroke that separated the feuilleton from the news, would theatre, book, and film reviews brush up against descriptions of political rallies, sociological and philosophical observations, and travel reports. Where [else] would all of those share space with short stories and serialized novels? (Lovett, 2017: 4–5)

Benjamin's adoption of the feuilleton fragment as key practice for his literary and political ambitions, and his gathering of those pieces together in new forms of collaged narrative producing sequences representing divergent thoughts over a period of time, provides us with an understanding of fragmented narrative practice. The impulses underlying it are:

- to replicate processes of our thinking and the ways we communicate – not linear and streamlined, more often random and piecemeal;
- to represent the world as we actually perceive it – complex, contradictory, disjointed, sometimes baffling;
- to see and judge the world politically – the author wanting to pass on the benefit of their intense experience and their analysis garnered from a bewildering array of sources and pressures perceived in society; and
- to investigate writing itself – to test the usefulness of those established techniques and conventional genres based on convergent thinking, in the context of development in technologies, politics, morals and human interaction.

Later in the century, other theorists followed Benjamin with influential fragmentary works: for example, Theodor Adorno's *Minima Moralia: Reflections from a Damaged Life* (1951), Barthes' *A Lover's Discourse: Fragments* (1978 [1977]), Rachel Blau DuPlessis' *The Pink Guitar: Writing as Feminist Practice* (1990) and Jean Baudrillard's *Fragments: Cool Memories III, 1990–1995* (1997). Brilliant memoirist social commentators such as Fernando Pessoa (*The Book of Disquiet*, published posthumously 1982), Sven Lindqvist (various from the 1960s onwards), and Svetlana Alexievich (*The Unwomanly Face of War*, 1985) wrote down fragmented thinking that produced powerful mosaicked narrative. Fiction writers on all continents have written in fragments in the hope of replicating the plural and unpredictable nature of real experience. And biographers, not feeling bound to represent the chronological flow of histories they focus on, have resorted to fragmentary representation of lifetimes because history comes to us actually as a mosaic of human recall and the availability of diverse extant documents (see, for example, Hughes-Hallett, 2013).

Lucy Hughes-Hallett's *The Pike. Gabrielle d'Annunzio: Poet, Seducer and Preacher of War* (2013) is a biography of the Italian novelist, playwright and poet Gabriele D'Annunzio, (1863–1938) who thrived in the turn-of-the-century and WW1 eras. It is written in fragments because, as she says:

> I have tried to avoid the falsification inevitable when a life – made up, as most lives are, of contiguous but unconnected strands – is blended to fit into a homogeneous narrative... Placing comments and anecdotes alongside each other like the tesserae in a pavement, *my aim has been to create an account which acknowledges the disjunctions and complexities of my subject, while gradually revealing its grand design...* Images and ideas recur in d'Annunzio's life and thought, moving from reality to fiction and back again: martyrdom and human sacrifice, amputated hands, the scent of lilac, Icarus and aeroplanes, the sweet vulnerability of babies, the superman who is half-beast, half-god. I have laid out the pieces: I have shown how they shift. (Hughes-Hallett, 2013: 16–17; my italics)

Hughes-Hallett takes her cue from evidence that d'Annunzio in 1896 delighted in helping a tiler lay a pavement in Venice. A recent visit that I took to d'Annunzio's extraordinary house, *Il Vittoriale* on Lake Garda, now kept as a museum, convincingly showed me that the author was obsessed with the collection and placement of myriad fragments in the pursuit of style and beauty in his life (see also Guerri, 2018). As Virginia Jewiss explains about D'Annunzio's fragmented novel, *Notturno*, written while he was convalescing from a WW1 injury to his eyes:

> To keep his sentences from overlapping and running together... d'Annunzio recorded his own brief thoughts... on thin strips of paper. Each wide enough for just one or two lines of writing, which his daughter Renata... prepared for him. (Jewiss, 2011: vii)

This method, enforced by temporary blindness, produced the mosaic focus, the extreme economy, and the read-between-the-lines qualities of the *Notturno* narrative:

> I write not on sand, I write on water.
>
> Every word I trace vanishes, as if abducted by a dark current.
>
> It is as if I can see the form of every syllable I record through the tips of my index and middle fingers.
>
> But only for an instant, accompanied by a glow, a sort of phosphorescence.
>
> Then the syllable dies out, disappears, lost in the fluid night.
>
> (D'Annunzio, 2011 [1921]: 13)

Clearly, writing in the dark affected the author's process – he replaced his signature baroque hyperbole and expansive melodrama with the compelling poetic insight available from the technique he was forced to use: the need to be concise, the staccato structure; the lack of opportunity to embellish at will. *Notturno* is 'the most emotionally direct and formally original' of d'Annunzio's prose works, earning even Hemingway's grudging admiration (Hughes-Hallett, 2013: 374). The immediacy of the writing throughout the novel derives from the strategy of bringing together thought, image and emotion in a constantly disrupted, fragmented narrative and allowing them to impact without explanatory ornamentation.

Experimental writers later in the 20th century continued to explore the concept of mosaicked narrative structure developed by Benjamin and d'Annunzio. There is a very long list of creative writers whose works employ discontinuity and fragmentation to provoke the reader into joining in on the meaning-making by thinking divergently and interpreting the gaps. I do not here canvass the multitude of poets who expressed divergent thinking and relied on it in readers to understand their work. Instead I indicate just some of the prose writers who required the reader's mosaic thinking to grasp what their experimental works were doing: John Dos Passos in his *Manhattan Transfer* (1925) and *USA* trilogy (1930–1936); William Faulkner in *The Sound and the Fury* (1929), *As I Lay Dying* (1930) and *The Wild Palms* (1939); John Steinbeck in *Cannery Row* (1945); Marc Saporta with *Composition No. 1* (1962); B.S. Johnson's novels in the 1960s; Julio Cortázar's *Rayuela* (1963) (trans. *Hopscotch*, 1966); several works by Italo Calvino in the 1970s; Sven Lindqvist's *'Exterminate All the Brutes'* (1992) and other novels; Toni Morrison's *Paradise* (1997); and Jonathan Safran Foer's *Extremely Loud & Incredibly Close* (2005). This list names just an acclaimed few of the prose writers who stuck their necks out against conventional thinking in their readers, and who sought instead a divergent mode of reading. Speaking about critical reaction to the fragmentation technique of *Paradise* in 1998, Morrison said:

> People's anticipation now more than ever for linear, chronological stories is intense because that's the way narrative is revealed in TV and movies... But we experience life as the present moment, the anticipation of the future, and a lot of slices of the past. (Morrison, quoted in Mulrine, 1998: 22)

In saying this, Morrison indicated that switched-on authors of the late 20th century still thought about writing in fragments in the same terms that writers used at the beginning of the century. This kind of writing cuts through the streamlining/editing of linear-text convergent thinking superimposed upon narrative; it represents more truly the divergent thinking processes occasioned by human experience and interpretation of the world.

An example to focus on here is Sven Lindqvist's *Exterminate All the Brutes* (1996 [1992]). In this memoir, Lindqvist stabbed facts and experiences down, pushed them around, bumped them up against each other, wove them inexorably, and seduced the reader. It's as if chronology – the supposed basis of narrative flow – was discarded, left to the reader to engage with and self-provide on an individual basis. The logic of the argument was built from strategically arranged contributing slivers. The gaps between the slivers were crafted to create thought processes that provide the overall argumentative verbal thinking, the visual thinking, and the narrative structural thinking that go together to create the full immersive experience for the reader. One might surmise that the notes for the writing were pushed to become the writing itself. These notes were highly crafted, edited down, placed, and polished to perfection, but they were still recognisable as divergently assembled. By page 10 of *Exterminate All the Brutes*, 13 fragments have been narrated. The staccato montage and stacking of narrative elements sets up a compelling effect for the argument. I will quote just three of the fragments:

1
You already know enough. So do I. It is not knowledge we lack. What is missing is the courage to understand what we know and to draw conclusions.

3
The core of European thought? Yes, there is one sentence, a short simple sentence, only a few words, summing up the history of our continent, our humanity, our biosphere, from Holocene to Holocaust.

It says nothing about Europe as the original home on earth of humanism, democracy, and welfare... [The sentence Lindqvist refers to is that of the title of his book: Conrad's 'Exterminate all the brutes'.]

7

The sound of heavy blows from a club, falling on the larynx. A crackling sound like eggshells, then a gurgling when they desperately try to get some air...

(Lindqvist, 1996: 2, 3, 6)

Lindqvist made an overall mosaic out of a sequential montage. It involved not only the splicing together of different moments in time but, also, the overlapping/superimposing of different voices, perspectives, genres, emotions and arguments to provoke the reader into discovering associations and realising new insights. In the process, the writer was aware of readerly potential for putting images and ideas together to make meaning happen, and intentionally directed the thinking of the reader by the juxtaposition of the fragments in a selective manner.

In the 21st century, creative writers have sought to write in a manner that combines experimental findings from the previous century with the impact of further technological advance. Jonathan Safran Foer, Jennifer Egan, Shelley Jackson and others recognised the influence of the internet and visual culture and how these entered the space once occupied exclusively by text, especially via the multimodal and divergent thinking these forms promote. Creative writers now need to look sideways and longways because the impacts from media require that fragmentation and divergent thinking are more an issue than they were before. The novel, the memoir, the essay and the short story, in their traditionally published convergent forms, still thrive. However, their linearity is impacted now not only by visual and audio genres playing in their space but also by the fragmentation inherent in contemporary sites of publication: the computer screen, hypertext and hypermedia.

Multimodal reading and writing is the norm in the 21st century. As I said in *Creative Writing and the Radical* (2016a), when citing Gunther Kress's analysis of multimodal literacy:

the way we read has changed due to our exposure to screen-driven culture. We now interact with diverse incoming channels [from our senses] to produce 'a *rich* orchestration of meaning' [Kress, 2011: 05.05]. Not only is the visual more pervasive in communication, but also, we are much more accepting of the idea that visual and written (or spoken) texts will operate in unison... Kress talks about finding meaning 'where it is' [Kress, 2011: 05.18]. In other words, we read and understand by *synthesizing* the variety of input modes that come in to us: by selecting, arranging and interpreting what we find most salient amongst the variety of incoming channels [from which] we form our reading, our created 'text'. (Krauth, 2016a: 12–13; my italics)

Five years ago, I did not realise just how rich Kress's (2011) 'rich orchestration of meaning' might be, and just how much synthesising mind-work was required not only to read a text but also to write it. Reading and writing multimodally involve all the associative thinking that William James analysed – the convergent or divergent linking of thoughts that might be triggered by the full range of senses and the emotions as well (James, W., 2019: 269). In this situation, the creative writer needs to know how to control, at least to some extent, the meaning-making thought process.

'Minding the Gap': The Mosaic Work of Writers

The concept that a piece of writing might be fashioned in such a way as to hand over a significant part of its meaning-making to the reader was picked up by literary critic J. Hillis Miller in his article 'The ethics of hypertext' (1995):

[I]n the period now coming to an end when the printed book dominated as the chief means of storing and retrieving information, it was still possible to be beguiled into thinking of a work like [Trollope's] *Ayala's Angel* or even like [Proust's] *A la recherche du temps perdu* as a stable and unmoving organic unity, on the model of a spatial array. Such a fixed text imposed on its readers a single unified meaning generated by a linear reading from the first word through to the end, in Proust's case more than three thousand pages later. The reader who accepted this model could think of the act of reading as a purely cognitive matter. I as reader do not create a meaning that did not exist before I actively engaged myself, 'interactively,' in the text. The meaning was there, waiting to be generated in me in an act of essentially passive reception. A hypertext that is overtly organized as such, on the other hand, offers the reader the necessity at every turn of choosing which path to follow through the text, or of letting chance choose for him or her. Nor is there any 'right' choice, that is, one justified objectively, by a pre-existing meaning. A hypertext demands that we choose at every turn and take responsibility for our choices. (Hillis Miller, 1995: 38)

Hillis Miller saw, early on in the development of hypertext, that it laid bare how readers really read – they choose paths forward amongst the array of fragments (words, sentences, chapters, images, feelings, etc.)

offered to them by writers – and especially how concepts of linearity and sequentialism are at odds with a text where a writer 'overtly organizes' it to be interactively manipulated by the reader. Prior to this, much argument was generated by Wolfgang Iser, Stanley Fish, Roland Barthes and other reader-response theorists on the role of the reader in interpreting the linear text, but, in my reading, less attention has focused on *the creative writer's willing understanding* that the reader's response will contribute significantly to the work. In my view, creative writers working with fragmented texts have always understood that the reader is a co-creator and they have written thoughtfully towards that outcome across a range of the work's functions. The idea for creative writers is that we place the tiles for an effective mosaic but we fully understand that readers may see it as a pile of shards. In that case, as with Coover's 'The Babysitter', we hope the reader will create a mosaic of their own whose overall meanings correlate with the one we intended. Or, as with Marc Saporta's *Composition No. 1* (1962) – a novel published on shuffleable pages – we present the pile of shards for the readers to do with as they will. In all this, there is the creative writer's acceptance of *multiple* readings. Any writer who thinks they will create in their readers only a singular intended meaning is way out of date. Writers know today their writing must cope with readings of all sorts, and they must seek to pre-empt them by thinking through as many divergent reading outcomes as possible, and by attempting to control them.

William James explained filling-in-the-blanks thinking (i.e. the cognitive aspects of reader-response theory) in terms of 'a fringe of felt affinity' that occurs with 'a gap we cannot yet fill with a definite picture, word, or phrase' (James, W., 2019: 260, 259):

> Whatever may be the images and phrases that pass before us, we feel their relation to this aching gap. To fill it up is our thoughts' destiny. Some bring us nearer to that consummation. Some the gap negates as quite irrelevant. Each swims in a felt fringe of relations of which the aforesaid gap is the term. (James, W., 2019: 259)

Iser's classic studies in reader-response theory (Iser, 1971, 1972, 1978) described 'the blank' in a text as not only a seemingly negative 'indeterminacy' but also a potential connection:

> the blank... designates a vacancy in the overall system of the text, the filling of which brings about an interaction of textual patterns... [and] the need for combination. [The blanks] indicate that the different segments of the text *are* to be connected, even though the text itself does not say so. (Iser, 1978: 182–183; italics in original)

Iser acknowledged the challenge upon the thinking reader but he also evoked the challenge upon the writer, to deal with the divergency of thinking. Iser examined how the reader of a conventional text picks up on and interprets the 'unseen' junctures that trigger meaning. He proposed that the reader is already 'programmed' (Suleiman, 1980: 24) to fill in the gaps in the text created by the author. William James and other psychologists provided the tools to understand this filling in of gaps: the stream of consciousness always requires the putting together of distinct and diverse thoughts but, while in logical convergent thinking the 'gaps' may be minimised and easy to negotiate, in divergent thinking more mind-work is required. A key quote from Iser:

> [T]he text is constructed in such a way that it provokes the reader constantly to supplement what he [sic] is reading... Whenever this occurs, it is clear that the author is *not* mobilizing his reader because he himself cannot finish off the work he has started: *his motive is to bring about an intensified participation* which will compel the reader to be that much more aware of the intention of the text. (Iser, 1971: 33; my italics)

Here Iser allowed the idea that the writer 'mobilizes' the reader to participate in the meaning-making, and he suggests that the author *could* finish off the work by applying strict convergent thinking methods but chooses not to. Reader-response critics such as Susan R. Suleiman (1980) soon pointed out that Iser contradicted himself over the intentionality of the text/author and the nature of the response the reader makes. Suleiman noted that:

> According to Iser, it is because all texts contain elements of indeterminacy, or 'gaps,' that the reader's activity must be creative: in seeking to fill in the textual gaps – gaps that function on multiple levels, including the semantic level – the reader realizes the work. But here again, the question of how much freedom the reader has is... answered in contradictory ways. Iser's conclusion is that 'the literary text makes no objectively real demands on its readers, it opens up a freedom that everyone can interpret in his own way' [Iser, 1971: 44]. This conclusion is opposed, however, by the weight of numerous other statements which suggest that the reader's activity of filling in the gaps is 'programmed' by the text itself, so that the kind of pattern the reader creates for the text is foreseen and intended by the author. (Suleiman, 1980: 24–25)

The concept of authorial intention as interpreted by critics is clearly of importance to the interests of creative writers. It is debated famously in the work of Roman Jakobson, Wayne C. Booth, Roland Barthes and others, but seeing it debated among reader- and audience-response theorists is ironic. Author intention is an idea constantly shied away from in audience-response theory, since the theory is predicated on the *reader* being responsible for the reading. Reader-response theorists are more comfortable with an 'implied author', rather than a real one. Suleiman calls both reader and writer 'necessary fictions' (Suleiman, 1980: 11).

Without denying the significance of the debate among audience-response theorists over the status of the 'gap' or 'blank', it can be said that where gaps in a narrative are conventional (e.g. between words, between sentences and paragraphs, between section and chapter breaks) readers in effect do not realise they are reading gaps at all because they are hardwired by familiarity to thinking that they are reading an uninterrupted convergent narrative, for which they supply the connections easily via convergent thinking. In written works where fragments and gaps are brought deliberately to the reader's attention and are fashioned by the writer *beyond* accordance with convention, the writer manipulates the reader more, drawing them into using their divergent thinking. The writer is not ignorant of the gaps, nor does s/he rely on conventional linear assumptions for their interpretation. The writer stage-manages these gaps, attempts to build *for-purpose* significance into them, and shapes them to direct reader understanding and make meaning. In a real sense, by creating a gap – or possibly filling it with an asterisk or some other symbol to represent a gap – the writer writes nothing at all, or writes a sign representing nothing, and *that* nothing is meant to be pregnant with meaning. In creating such a device, the writer considers the impact, the drama, the interpretability and ideational likelihood of the 'missing' text. The writer *writes by not writing text*, by consciously refraining from filling/ explaining/forcing meaning into the gap, but by intending meaning nevertheless. The writer thinks deeply about the gap and intends by textual absence potential meaningful thinking for the reader to engage in.

In the gap, the writer expects thinking process will happen. The gap between unlike thoughts is, as William James says, 'aching' to be filled – at least in those minds that are up for the challenge of reading the gap. The contemporary writer intends that there will be readable frisson between the fragments, that textual energies and divergent thinking will flow, that narrative dynamics will emerge to relate fragment to fragment near and far within the work, and to relate groupings of fragments to the whole. Themes, plots, whole characters or theses will emerge from the interplay allowed in the gaps. Each reader might get a different version of the themes, plots, characters or overall expository outcome, but that was going to happen anyway in a conventional work requiring conventional reading. The point here is that the writer's process involves directing the reader's negotiation of the gap and guiding the reading of *no text* in ways not dissimilar to how the writer naturally attempts to control the reader by conventional text: it is being done by appeal to how people think, whether in convergent or divergent modes. The reader is invited, in reading the gap, to put a sequence of *no-text* + *adjacent texts* together, as placed by the writer, in a manner consistent with how they 'read' their fragmentary experience of the real world. There is an honesty – and a sense of reality – embedded in the idea of writing with fragments and celebrating the potential of the gaps; by this method, the writer taps into the reader's normal thinking and understanding processes, albeit with a little extra mind-work involved.

I can say that when I write in fragmented narrative mode (see, for example, Krauth, 2017, 2016b, 2015, or earlier, Krauth 2000, 1997) I set out, from the start, with the intention of utilising and manipulating the gaps between the sections of my narrative. It excites me, as writer, to bump fragments up against each other, to work with the electrical charge generated by juxtaposed discrete ideas and experiences. For me, it's like atoms or continents colliding. I recall writing in the 1990s a story where I set out to compose its pieces in the unplanned order they came to me in, with a promise to myself that I must rearrange them later on, at the time of editing, to make sure the final version made sense. In the end I found, very surprised, that I could not better the order in which they first arrived in my head. The fragmented story was published in exactly the sequence I wrote it down (Krauth, 1996). This method – perhaps to be called 'placing it anyhow', or 'smashing it out', or 'picking up the next tile and placing it seemingly regardless' – I discovered, was a dependable way of writing creatively, because the directions for reading the fragments were already coded by the way my mind's divergent thinking came up with the sequence. And I realised that 'minding' the gap between fragments does not mean paying focused and rational attention to them: it means allowing *my mind* to do the organising – that kind of *minding*. When we read, our brains put the bits together subjectively, even when the mode is linear. We (or at least those who aren't brainwashed) are not so hard-wired that we think only linearly as we read. All readers acknowledge, for example, that when we read a text a second time we see, hear and feel it differently. Sometimes we take a *different* route of thinking to a conclusion previously reached; often we do not: our minds take us on another route entirely to somewhere else, a new conclusion. The creative writer who is happy to invite multiple conclusions to the reading of their work, as I am, perfectly accepts this situation. I think the writer should trust that the minding of the gap that their own brain does will significantly correlate with the minding the reader does ... because 'minding' – the making of meaning in the mind – does not involve just convergent thinking but also divergent thinking.

The Wandering Mind: Writing and Fantasising

One of the great 'interferers' in our convergent thinking is mind-wandering. Scientists suggest that mind-wandering, while it may be disruptive, also plays a major role in creativity and invention:

> The brain itself seems to be a natural source of random variation, mediated largely through the default-mode network [i.e. the default brain activity that occurs when we are supposedly 'at rest'], and perhaps most evident in our dreams – although our nightly dreams are mostly forgotten and therefore only weakly available for selective retention. However, mind-wandering occupies nearly 50% of our waking hours... and is perhaps the most available source of creative ideas. (Corballis, 2018: 55)

Famous 'mind-wanderers' have included scientists, architects and inventors such as Albert Einstein, Frank Lloyd Wright and Nikola Tesla along with writers Samuel Taylor Coleridge, Robert Frost and Virginia Woolf (Carson, 2018: 140). Today we might recognise 'mind-wandering' in the form of ADHD (attention deficit hyperactivity disorder) in ourselves or those around us. Carson reported that a national study in Sweden 'found higher rates of ADHD in members of the writing population than in nonwriters in the Swedish population' and that children and adolescents with ADHD tend to record 'increased scores on aspects of creativity or divergent thinking tasks' (Carson, 2018: 140).

Our attention can be diverted towards externally or internally presented stimuli – things going on around us, or things going on in our heads. Cognitive psychologist Mathias Benedek noted that 'cognitive science has traditionally focused on the study of externally directed cognition', but that more recent studies:

> suggest that [while we] we spend 25%-50% of our waking time engaged in thoughts unrelated to what is going on around us... [d]uring these states our attention is directed to internal trains of thought that are commonly concerned with past experiences or future plans... These stimulus-independent thoughts can be spontaneous or goal-directed. Spontaneous stimulus-independent thought includes the phenomenon of mind-wandering, where

thoughts are unintentionally drawn away from a task... In contrast, goal-directed stimulus-independent thought includes activities like deliberate planning or idea generation. (Benedek, 2018: 180)

Thus mind-wandering can be as useful as it is distracting. Cognitive psychologists Jonathan Smallwood and Jonathan W. Schooler studied individuals with wandering minds and found:

Mind wandering... involves a complex balance of costs and benefits: Its association with various kinds of error underlines its cost, whereas its relationship to creativity and future planning suggest its potential value. (Smallwood & Schooler, 2014: 487)

Smallwood and Schooler say mind-wandering is 'essential to the stream of consciousness':

Conscious experience is fluid; it rarely remains on one topic for an extended period without deviation... Studies exploring the phenomenology of mind wandering highlight the importance of its content and relation to meta-cognition in determining its functional outcomes. Examination of the information-processing demands of the mind-wandering state suggests that it involves perceptual decoupling to escape the constraints of the moment, its content arises from episodic and affective processes, and its regulation relies on executive control. (Smallwood & Schooler, 2014: 487)

So, mind-wandering involves 'perceptual decoupling': looking at other things, maybe, or looking inward rather than outward. It involves 'episodic and affective processes' such as disruptive memory or intrusive emotional thinking. It also involves metacognition – being aware of 'information from memory', for example (Smallwood & Schooler, 2014: 505). But while the day-dreaming version of metacognition may not be directed at *analysing* how our mind is operating, it does involve us *paying attention* to our conscious thinking. We are not generally self-consciously alert to our thoughts when we operate normally, but we are certainly aware of them when we fantasise, and especially so in the moment after we 'snap out of it'. Finally, Smallwood and Schooler mention that the 'regulation' of mind-wandering 'relies on executive control' – *executive control* being the brain's ability to manage its cognitive processes. But scientists are undecided about how mind-wandering can involve both poorer and greater executive control. Writers are familiar with both sorts of mind-wandering – those by which we lose track of what we are doing, and those that forge new tracks to new ideas – and this is the nub of the issue. Smallwood and Schooler (2014) concluded: 'Aware versus unaware mind-wandering episodes differ in the situations in which they are most likely to arise as well as in their respective impact on performance and brain activity' (Smallwood & Schooler, 2014: 510). Thus the brain's executive function may continue and enhance the wandering, may fail to stop it happening or, indeed, may call a stop to it. Fiction writers manipulate mind-wandering on purpose to find the next scene, the next action, the next character thought, just as essay writers use it at times to find the next persuasive thing to say.

Sadly, for creative writers, Freud did not devote much time to the creative writer's mind, even though he used literary works on many occasions in his studies. However, when he did write about creative writing, he used the context of day-dreaming, i.e. mind-wandering, as entry point. Freud's unfinished essay 'Creative writers and day-dreaming' (1908) begins with a beautiful paragraph of admiration for writers from the point of view of those who enjoy reading:

> We laymen have always been intensely curious to know… from what sources that strange being, the creative writer, draws his [sic] material, and how he manages to make such an impression on us with it and to arouse in us emotions of which, perhaps, we had not even thought ourselves capable. Our interest is only heightened the more by the fact that, if we ask him, the writer himself gives us no explanation, or none that is satisfactory; and it is not at all weakened by our knowledge that not even the clearest insight into the determinants of his choice of material and into the nature of the art of creating imaginative form will ever help to make creative writers of *us*. (Freud, 1959 [1908]: 420)

Proceeding with the view that *the writer himself gives us no explanation*, Freud supposed that to find the secret to being a creative writer, he should examine childhood and the fantasising play undertaken there:

> Might we not say that every child at play behaves like a creative writer, in that he creates a world of this own, or rather, rearranges the things of his world in a new way which pleases him? (Freud, 1959: 421)

Developing his argument, Freud said that, in adulthood, everyone swaps youthful play for more mature day-dreaming – 'according to the sex, character and circumstances of the person who is having the fantasy' (Freud, 1959: 423) – and in writers of genre-level novels, romances and short stories, 'who nevertheless have the widest and most eager circle of readers of both sexes' (Freud, 1959: 425), the hero-orientation of their work reveals 'His Majesty the Ego' in ascendancy. This, he says, is reflected too in more literary writing:

> It has struck me that in many of what are known as 'psychological' novels only one person – once again the hero – is described from within. The author sits inside his [sic] mind, as it were, and looks at the other characters from outside. The psychological novel in general no doubt owes its special nature to the inclination of the modern writer to split up his ego, by self-observation, into many part-egos, and, in consequence, to personify the conflicting currents of his own mental life in several heroes. (Freud, 1959: 426)

It is a pity that Freud's argument ran out of gas around about the time he said: 'a piece of creative writing, like a day-dream, is a continuation of, and a substitute for, what was once the play of childhood' (Freud, 1959: 427). Freud acknowledged that 'when a creative writer presents [their work] to us or tells us what we are inclined to take to be his personal day-dreams, we experience a great pleasure' (Freud, 1959: 426–427), but he ended the essay with: 'How the writer accomplishes this is his innermost secret... This brings us to the threshold of new, interesting and complicated enquiries; but also at least for the moment, to the end of our discussion' (Freud, 1959: 428). If only Freud had come back to this topic, we might *possibly* now better understand how writers' minds work – at least in Freudian terms.

The rich collage/dynamic/fabric/space/landscape of our minds comprises many interrelated parts, and a significant portion of our constant thinking involves managing current experience, memories and fantasies. The mind is a complex system, and is itself a mosaic in many ways, but also it must be a focusing mechanism: our ordinary lives depend on us getting our thoughts in order. Creative writing, too, requires the writer to deal with the kaleidoscopic mosaic of mind-activity, and some of that involves finding straighter paths through the puzzle of thinking – ignoring the mosaic mind, and seeking access to the flowing mind.

5 The Flow Mind: Writing and Convergent Thinking

The Flow Mind

Research about creativity considers the idea of the mind in 'flow'. Psychologist Mihaly Csikszentmihalyi's influential flow theory proposes that the individual engrossed in a task can enter a zone of enjoyable forward progress based on 'the feeling' that things are 'going well as an almost automatic, effortless, yet highly focused state of consciousness' (Csikszentmihalyi, 1996: 110). According to Csikszentmihalyi's *Creativity: Flow and the Psychology of Discovery and Invention* (1996), the flow zone involves nine main elements:

(1) There are clear goals every step of the way.
(2) There is immediate feedback to one's actions.
(3) There is a balance between challenges and skills.
(4) Action and awareness are merged.
(5) Distractions are excluded from consciousness.
(6) There is no worry of failure.
(7) Self-consciousness disappears.
(8) The sense of time becomes distorted.
(9) The activity becomes autotelic [i.e. worth doing for the sake of the experience itself].

(adapted from Csikszentmihalyi, 1996: 111–113)

A writer perusing these characteristics will probably shout: *Bring it on!* In reading Csikszentmihalyi, I presume a similar reaction came from the 'chess players, rock climbers, dancers and composers... athletes, artists, religious mystics, scientists, and ordinary working people' who took part in his experiments (1996: 110). The achievement of such exceptional process quality cannot be an everyday occurrence.

From experience, creative writers know that creative projects typically start out stutteringly, and that none of Csikszentmihalyi's nine elements operates from the very outset. For example, regarding (1), initially there may be no apparent way forward, just dimly perceived clues, and, the further we grope our way, the clearer the goal might become. Or with (2), at most times in completing an individual project, there is likely only feedback generated by the practitioner themself that can usefully apply. Or with (3), the balance between the challenge and the skills changes regularly as the project continues and may remain uncertain until the end of the activity. Elements (4) – (9) are similarly questionable.

Csikszentmihalyi's flow concept describes a point at which the hard work is already done to get the mind and body into a zone of seemingly effortless output. In his 'Conditions for flow in creativity' section (1996: 113–123), he admits that 'for artists the goal of the activity is not so easily found. In fact, the more creative the problem, the less clear it is what needs to be done' (1996: 113–114). Csikszentmihalyi then appeals to the idea that 'the creative person somehow must develop an unconscious mechanism that tells him or her what to do', and goes on to say, 'the ancients called that voice the Muse' (1996: 114–115). Using novelist Robertson Davies' example of a recurring mind-image which prompted the writing of a novel, Csikszentmihalyi says: 'Very often this is how the Muse communicates – through a glass darkly, as it were. It is a splendid arrangement, for if the artist were not tricked by the mystery, he or she might never venture into the unexplored territory' (1996: 115). With these statements, to my mind Csikszentmihalyi no longer describes anything he knows about; he resorts to speculation and myth, not seemingly the stuff of Psychology.

I don't have space to deal in detail with Csikszentmihalyi's conditions for all nine of his flow elements. As a writer I say that *none* of them is convincing because each is based in the idea of creativity as *convergent* problem-solving rather than *divergent* exploration. Csikszentmihalyi's flow theory, while aimed at imbuing practitioners with a means of joyful confidence, fails to deal with issues the writer constantly faces, such as:

(1) When does the creative writer ever know exactly where they are going? Even when we put the full stop to a major work, we likely know there are things that could, or should, be worked on further.
(2) How do we ever know how to treat feedback? Those who read our manuscripts have differing (sometimes wildly differing) opinions, as do those who read our published work.
(3) Individuals regularly have difficulty assessing their own capabilities. How do we ever know that we are up to the task when others (such as critics and competition judges) so often claim the right to decide that for us?

With reference to Csikszentmihalyi's elements (4) to (9) – where the creative practitioner is really *flying* – I have had just two such experiences in a lifetime of writing. On those occasions I achieved a kind of orgasmic transcendence while writing a novel. On the afternoons in question, I had been writing all day and worked myself to a narrative facility where everything flowed from my mind with perfect choice of words and sentences, perfect voice and pitch, and perfectly captured viewpoint, energy and drama. Those two afternoons (in 40 years of endeavour) produced work where not a single act of editing was required; the final manuscripts (two passages of approximately 500 words each) came out perfectly formed for publication. This was Csikszentmihalyi's flow elements (4) – (9) in action. For me, those occasions were epiphanies in writing; but such experiences hardly ever happen.

Transcendent moments cannot constitute a writing process method. They indicate exceptions rather than rules. So, we have to find ways to approximate those best-ever times when the mind at a high level of organisation flows unimpeded onto the page. Csikszentmihalyi in effect talks about the joy of creation once the problem of what is being created is overcome and the goal is clear. For writers, this may occur when we believe we know what we are doing to complete a story, poem, novel, play, etc. But while Csikszentmihalyi's flow, being about sequential thinking and events involving problem-solving and argument-making, *might* work better for essayists, journalists and scientists, it is not a solve-all for creative writers. The creative writer is mostly involved in generating possibilities and following them through to conclusions that are 'logical' only in terms of the character, situation, plot and setting parameters established overall for the piece. In fiction, poetry and plays, the 'logical' outcome to an all too human character development or plot situation will always be subject to accidents, complications, coincidences, distractions, sub-plots, minor events, red herrings, etc. (thus replicating life), and will likely represent another kind of flow entirely in the narrative outcome. The special logics that apply to creative works do not easily fit with the scientific problem-solving logics that Csikszentmihalyi espouses.

A case in point is when mosaic thinking (see previous chapter) really gets going and the fragments contributing to the narrative line up promisingly: where the creative writer works out the unconventional patterning between the fragments, and – on a high, such as Csikszentmihalyi describes – just keeps going. In this instance, the joyful realisation is not of a '*clear* goal every step of the way', but of immersion in a process where a dynamic fluidity (not a standard logic-driven evolvement) looks and feels great on the page and makes a strong foundation for editing that will follow. This whirling vortex of mind-material funnelling into a written product, albeit with warts attached, is a good example of the high that creative writers seek. It is rather like the difference between *catching one train directly to the destination* (as Csikszentmihalyi seems to describe) or *changing trains many times to get there just the same* (as I might put it). The latter way of getting there may take longer and may involve doubling back, repeating legs of the journey, making unpromising snap decisions, etc., but if there are good enough connections at each change station, then the process is a success.

Csikszentmihalyi's theory can reflect William James' stream-of-consciousness when the carriages exiting the mind-depot line up ideally: logically coupled, riding the rails smoothly in convergent ways. All writers know that their messy drafts, whether divergently or convergently created, need significant convergent thinking to edit them into focus as publishable products. But forms of writing seen to involve from the outset convergent rather than divergent thinking (e.g. journalism, public writing, science writing) still require an understanding of how the flow mind and the mosaic mind fit together. Even editing – supposedly requiring the laser-focus of convergent thinking – can benefit from the insertion at various stages of newly divergent ideas.

Neurocognitive studies of flow have focused mainly on optimal perfor-mance of the body in physical activity because it is known that:

> engaging in simple forms of physical exercise induces significant changes in mental states, and is associated with several positive outcomes, includ-ing improved wellbeing and mood… Neuroscientific investigations have demonstrated that running, for instance, is associated with neurogenesis, neural plasticity and memory enhancement… (Abraham, 2018: 270)

For an understanding of why the flow state happens in the brain, Abraham suggests:

> The most dominant idea to date is that of 'transient hypofrontality'… In this framework, the state of flow arises when the frontal lobes of the brain, which are the key hub of the explicit system of the brain that orchestrate cognitive control, flexible problem solving, analytical and meta-conscious operations, are temporarily suppressed. The explicit system 'being offline' allows for the experiential and implicit system of the brain, which is facil-itated by subcortical brain structures like the basal ganglia, to take over. (Abraham, 2018: 256)

In other words, the brain's executive control goes on holiday to allow the flow to happen, but, according to Abraham, that includes the 'flexible problem solving' and metacognitive mechanisms taking a vacation, too. For writing, this suggests we should do our best work when we are going about it in a mindless way. For me, that just doesn't work. Somewhere along the line – and best done *all the way along the line* – the writer needs to work with and monitor their thinking, encouraging its flexibility while keeping control of its excesses. An afternoon of orgasmic writing is welcome but let's not rely on it happening often.

Clinical experiments that are focused on normal brain function during physical activity depend mainly on effects recorded before and after the activity is undertaken, because the brain is difficult to observe outside the laboratory. Creative flow performance in arts areas such as dance and music, where creative process and production are more physical than with writing (Abraham, 2018: 173–199), has received significant attention. The current findings can be adapted for relevance to creative writing. They suggest that mental changes occasioned by the writer getting up from their desk and doing something physical *other than* the minimal body movements used to generate writing will be beneficial. Also, they suggest that physical activity allows better access to a practitioner's memory store (Abraham, 2018: 187) and this can help get flow to happen.

Jono Lineen's book, *Perfect Motion: How Walking Makes Us Wiser* (2019), examines the flow mind in the context of writing a creatively structured memoir. The book is informed by the author's extraordinary life of walking in remote regions of the world. It follows not just his own memories and the findings of cognitive science but also the footsteps of acclaimed walker-writers from Rousseau to Cheryl Strayed. In the PhD submission underlying the book, Lineen says:

> Csikszentmihalyi's research affirmed for me that there was a universal state of mind, anyone can enter, that has the ability to generate original ideas. According to Csikszentmihalyi, flow is achieved when we are engaged in an activity we have mastery over and yet it still challenges us; this state of balance is known as the flow channel. Walking has a consistent ability to create flow. In fact, Csikszentmihalyi uses walking as a case study in the creation of flow states. He says: 'A good example of this method [of achieving flow] is the act of walking, which is as simple a use of the body as one can imagine, yet which can become a complex flow activity, almost an art form'… (Lineen, 2020: 269)

Achieving the flow mind by sitting almost stationary at the writing desk may not happen. But accessing flow via more challenging physical activity provides an entrée into better mind-work. Writers have reported regularly that walking puts them into a good initial frame of mind for writing, and that this frame of mind flows on into all stages of the writing process, whether it be when we are thinking about how to continue what we have already written, or thinking about how to improve what we have completed. Thus, taking a walk *during* writing sessions can be beneficial, too.

Csikszentmihalyi (1996) talks about the 'dry spells' a creative writer confronts – created often by writer's block – and admits that 'total immersion in the flow of writing… cannot be sustained for long periods of time' (Csikszentmihalyi, 1996: 242). He observes that a creative writer's modus operandi cannot be flow, flow, flow all the time, but consists of 'a constant alternation between a highly concentrated critical assessment and a relaxed, receptive, nonjudgmental openness to experience' (1996: 242). Because creative writing is such a difficult thing to do, a creative writer's mind benefits from fallow periods – time is needed for the ground to recover and prepare itself, for the new crop to begin to grow; a break is needed so that the batteries can recharge. In my experience, periods of writer's block are indeed difficult to deal with, but they are never resolved by focusing on the writing problem itself. Diverting myself, with a subtle metacognitive understanding that the diversion is really part of my writing process, has worked for me.

Getting the mind to flow, even if not in the ideal terms that Csikszentmihalyi describes, but at least with basic effectiveness in the writing situation, can be further explored with reference to walking and writing, memory and writing, substance intake and writing, and the idea of the social mind, as follows.

Writing, Walking and the Convergent Mind

Walking and thinking have long been linked. While it may be a myth that Aristotle's philosophy school walked while they cogitated – the name 'Peripatetic school' may come merely from the fact that they sat in the *peripatoi* [walkways] of the Lyceum in Athens – nevertheless, as Rebecca Solnit notes: '[following the ancients] the relationship between walking and philosophizing became so widespread that central Europe has places named after it' (Solnit, 2001: 16). Such walkways for thinking include the Philosophenweg in Heidelberg and the Philosopher's Way in Copenhagen. In her book, *Wanderlust* (2000), Solnit evokes the habits of a significant list of eminent walker–thinkers, including 17th-century English philosopher, Thomas Hobbes, who 'had a walking stick with an inkhorn built into it so he could jot down ideas as he went' (Solnit, 2001 [2000]: 16) – an early form of the smart phone.

Following the philosophers, great creative writing minds advocated the relationship between walking and the thinking required for writing. These included Jean-Jacques Rousseau, William Wordsworth, Henry David Thoreau, Charles Baudelaire, Virginia Woolf, Walter Benjamin and numerous others to the present. Rousseau, Wordsworth and Thoreau in particular promoted rural walking for calm reflection to achieve an unhindered mental flow where analysis of the world and existence could proceed. They suggested that escape from the pressures and distractions of the city, its society and politics, allowed the mind to focus on interiority and metacognition, giving it freedom to drift creatively but also to move purposefully. Walking inspired divergent *and* convergent thinking; as Rousseau (2008) said, it is both a means of 'release' and 'daring', and a channel for 'choosing' and 'combining':

> There is something about walking that animates and activates my ideas; I can hardly think at all when I am still; my body must move if my mind is to do the same. The pleasant sights of the countryside, the unfolding scene, the good air, a good appetite, the sense of well-being that returns as I walk… all of this releases my soul, encourages more daring flights of thought, impels me, as it were, into the immensity of beings, which I can choose from, appropriate, and combine exactly as I wish. (Rousseau, 2008 [1782]: 158)

On the other hand, Woolf and Benjamin walked in built environments and found ways to synthesise the disruptive and challenging conditions of modern existence into a mind-flow that also produced writing. Woolf said, of walking in Cambridge and Oxford:

> Strolling through those [Oxbridge] colleges past those ancient halls the roughness of the present seemed smoothed away; the body seemed contained in a miraculous glass cabinet through which no sound could penetrate, and the mind, freed from any contact with facts… was at liberty to settle down upon whatever meditation was in harmony with the moment. (Woolf, 1989 [1929]: 6)

The effects are almost identical for Rousseau and Woolf – freedom to think more widely, access to a focused mind-space – but the settings are very different: an isolated island *versus* a busy university town. The common factor is the walking.

The rural mind and walking

Jean-Jacques Rousseau's classic walking and writing study, *Reveries of the Solitary Walker (Les Rêveries du promeneur solitaire)*, was written during the period 1776–1778 and published (after his death) in 1782

(Rousseau, 20011 [1782]). It comprised ten chapters with the titles 'First Walk', 'Second Walk', 'Third Walk' and so on up to 'Tenth Walk'. They formed a set of meditations on a variety of topics, all undertaken (he said) as he walked on the secluded Ile de St Pierre in Lake Biel, Switzerland, which was his retreat in the final years of his life, where he attempted to put his thoughts together. Rousseau suggested that he did his Ile de St Pierre walking to collect plant species for a botanical compendium he was writing, but he had already explained the relationship between thinking and writing that underlay the *Rêveries* in his *Confessions* (written in 1769), which were also published posthumously:

> Seated at my table, with my pen in my hand and my paper in front of me, I have never been able to achieve anything. It is when I am out walking among the rocks and the woods, it is at night, sleepless in my bed, that I write in my head. (Rousseau, 2008: 111)

Here, Rousseau says the benefit he got from walking was equivalent to that experienced lying in bed and thinking at night. Two hundred years before neuroscience, Rousseau observed similar mind-flow effects for walking and for the liminal asleep–awake space, in the writing process context.

According to neuroscientists, wave patterns for mind-flow recorded in the period between states of waking and sleeping are dominated by theta waves:

> Theta activity has a frequency of 3.5 to 7.5 Hz and is classed as 'slow' activity. It is seen in connection with creativity, intuition, daydreaming, and fantasizing and is a repository for memories, emotions, sensations. Theta waves are strong during internal focus, meditation, prayer, and spiritual awareness. It reflects the state between wakefulness and sleep and relates to the subconscious mind. (Neurohealth, 2021)

Ned Herrmann, in *Scientific American*, explains:

> A person who has taken time off from a task and begins to daydream is often in a theta brainwave state… Individuals who run outdoors often are in the state of mental relaxation that is slower than alpha and when in theta, they are prone to a flow of ideas. This can also occur in the shower or tub or even while shaving or brushing your hair. It is a state where tasks become so automatic that you can mentally disengage from them. The ideation that can take place during the theta state is often free flow and occurs without censorship or guilt. It is typically a very positive mental state. (Herrmann, 1997: np)

Following Rousseau, many creative writers noticed the state of mental relaxation that promoted the writing–thinking process – whether it be produced by walking, by the part-asleep–part-awake state lying down, or by some other method of mind release. Rousseau, in his 18th-century way of thinking the issue through, likely understood that the French word *rêverie* was related to the Latin word *vagari*, meaning 'to wander or roam about' (Goulbourne, 2011: xxi). The words themselves told the story that a reverie state is similar to a vague state, where the mind flows not in specifically chosen directions, but with more liberty within an indistinctly defined territory. When he walked, Rousseau jotted down notes for the ten chapters of *Rêveries* on 27 playing cards, now on display at the Neuchâtel Museum in Switzerland. He said in his *Confessions* that his 'passion for walking' (Rousseau, 2008: 53) derived from what he knew about his mind-flow: 'I can meditate only when walking… as soon as I stop, I can no longer think, for my mind moves only when my feet do' (2008: 400). Rousseau set up ideas about mind-flow for walker–writers to follow. Wordsworth, Thoreau and Woolf, among many others, acknowledged their debt to him.

Probably the best-known walker–writers in the 19th-century were Wordsworth, Thoreau and Baudelaire. Strongly influenced by Rousseau's writings and creative practices (Legouis, 2000), Wordsworth made Rousseau's Ile de Saint Pierre a particular destination on his famous early walking trip across the Alps in 1790 (Goulbourne, 2011: xxv). Wordsworth spent much of his life walking, especially in the English Lake District, in order to progress his writing. Thoreau told the story: 'When a traveller [visiting Wordsworth's house] asked Wordsworth's servant to show him her master's study, she answered: "Here is his library, but his study is out of doors"' (Thoreau, 1862: 659). It was a mantra for rural Romantic writers.

Wordsworth's poem about daffodils (published in 1807) is his most famous analysis of the relationship between walking, writing and the mind. The poem began with one of the poet's typical walks, seeking the mind-flow necessary for inspiration, along Ullswater lake which lay 12–15 miles across country from his home at Dove Cottage in Grasmere:

> I wandered lonely as a cloud
> That floats on high o'er vales and hills,
> When all at once I saw a crowd,
> A host, of golden daffodils…

(Wordsworth, 2002 [1807])

The poem might seem to be simplistically about daffodils, and what the natural environment offered the writer in terms of inspiration.

But, importantly, it was about the writer's mind-flow. When the poet returned home and switched modes from *walking + thinking* to *lying down + thinking* (both of them theta brain-wave states) he said:

> For oft, when on my couch I lie
> In vacant or in pensive mood,
> They flash upon that inward eye
> Which is the bliss of solitude...
>
> (Wordsworth, 2002 [1807])

The interaction between the minutely perceived and closely felt environmental experience, the recording and storing of that experience in the mind, and the return in later mind-work to analyse it, constitutes the writer's ecology. This process demands a mind-flow that frees up perception and analysis during the collecting-together period when the observations are being made, while it also brings perceptions, experiences, feelings and words into convergent focus at the time when the data collected is reviewed and put into writing.

Thoreau added to the Romantic analysis of walking and writing in his essay 'Walking', published in *The Atlantic Monthly* in 1862. As with Rousseau's 'Walks' and, later, Woolf's strolling that produced her lectures, Thoreau's essay performed the *outcome* of thinking undertaken while walking. It started with thoughts about walking, and was soon dealing with other philosophical, political and social issues, all of which were nevertheless inspired by walking in the vicinity of his local village of Concord in Massachusetts. Thoreau saw each writerly walk as 'a sort of crusade' (Thoreau, 1862: 657), part of the 'noble art' of walking that led to thinking (1862: 659). This kind of walking:

> has nothing in it akin to taking exercise, as it is called, as the sick take medicine at stated hours, – as the swinging of dumbbells or chairs; but is itself the enterprise and adventure of the day... you must walk like a camel which is said to be the only beast which ruminates when walking. (Thoreau, 1862: 659)

The walking that Thoreau advocated involved contemplative awareness of the mind's working but it also meant being in touch with one's whole body: 'In my walks I would fain return to my senses,' he said (1862: 659), and these were the senses of seeing, hearing, smelling, feeling and tasting that the brain needed to be in flowing touch with for productive life and work.

The Romantic writers did not sit, confronted by a blank page (and perhaps a blank mind) attempting to think about what they might write.

They trekked into the world and made something happen in their minds to write about. They sallied forth, seeing writing as the collecting of perceptions, ideas and recollections to be worked on and synthesised into understanding. They climbed mountains, trod shores of isolated lakes, strode environs surrounding their village homes, determinedly seeking inspirational, cognition-triggering material. When Thoreau (1862: 659) wrote: 'In my walks I would fain return to my senses', he directed us to the literal meaning of the phrase *return to my senses*, which evokes the mind's flow of interpretative and integrating processes fully in touch with its incoming sense channels. Experiencing the sensual world and making mental sense of it, is the perennially inspiring project for writers.

The city mind and walking

Walking the city in order to experience a productive mind-flow is different from the process of walking in the countryside. The city is man-made, a construct at odds with the natural environments that the early Romantics sought contact with. But modern Romantics such as Woolf and Benjamin wandered streets, arcades and city parks in search of mind-triggering experience to match what earlier writers found in wild, unspoiled nature. In general terms, we can account for this change in writerly perception by noting a paradise-lost perspective in literary thinking in the late 19th century: city living had encroached so much upon people's cognition that the urban landscape became the thinking norm. In Paris in the 1850s, Baudelaire pioneered the idea that writerly city walking produced insights into the human condition in a way that walks in the countryside previously achieved, but of course, the sights and sounds and lessons learnt were far less pleasant to contemplate. In part, Baudelaire inspired Benjamin to produce his great piece of walking–writing: *One-Way Street* (1928), set in the conglomerate European city that Benjamin kept as a metaphor in his mind, based on his living in Berlin, Marseille, Naples, Moscow, Paris and other metropolises. In London, Virginia Woolf walked regularly in Tavistock Square, Bloomsbury, for the sake of her writing and mental health, but she made direct use of walking in the city of 'Oxbridge' in composing her greatest piece of walking–writing, *A Room of One's Own* (1929).

In *A Room of One's Own*, Woolf analysed the mind-work she did in composing two lectures on the topic of women and fiction to be delivered at Newnham and Girton colleges, Cambridge, in October 1928. Instead of producing conventional lectures, which she said would need 'to come to a conclusion' (Woolf, 1989 [1929]: 3), she traced her mind-flow during the days of compositional thinking leading up to the lectures. Worried that her audience might be disappointed at not receiving a conventional lecture, 'in order to make some amends,' she said, 'I am going to develop in your presence as fully and freely as I can *the train of thought* which led me' to produce the lectures (Woolf, 1989: 4; my italics). She created what is now a highlight in the history of literature – her thesis that demanded for women fiction writers a dedicated writing space and an income. *A Room of One's Own* shows that Woolf's 'train of thought' was an activity not sequestered from the ordinary thinking she did throughout the days of preparation. Rather, she 'made it work in and out of [her] daily life' (1989: 4). To the audience she said: 'I give you my thoughts as they came to me' (1989: 6–7).

Woolf described the flow of phases in her compositional process. The first phase – undertaken in the university city while 'sitting on the banks of a river in thought' (1989: 5) – described initial writerly thinking as a kind of fishing in the flow of consciousness:

> Thought… had let its line down into the stream. It swayed, minute after minute, hither and thither among the reflections and the weeds, letting the water lift it and sink it, until – you know the little tug – the sudden conglomeration of an idea at the end of one's line: and then the cautious hauling of it in, and the careful laying of it out? Alas, laid on the grass how small, how insignificant this thought of mine looked… But however small it was, it had, nevertheless, the mysterious property of its kind – *put back into the mind* it became at once very exciting, and important… (Woolf, 1989: 5; my italics)

Woolf revealed not only her ability for metacognition but also her understanding of how a writer intervenes in their thinking in order to organise it. She recognised the potential of her initiating thought, then she 'put [it] back into the mind' so as to manipulate and develop it further:

> …it darted and sank, and flashed hither and thither, set up such a wash and tumult of ideas that it was impossible to sit still. It was thus that I found myself walking with extreme rapidity across a grass plot. (Woolf, 1989: 6)

The effect her thinking had on her body was remarkable. Many writers will recognise the energy she describes: it is released when we are sure we have had a great idea. Woolf was up and running with this inaugurating idea for her lecture about women and fiction. But immediately – Oh, irony! – she found that she was crossing a patch of university grass that only men were allowed to walk on. Her unconventional approach to how to write a lecture called into question all the established rules about how 'proper' (male-view) writing was done.

Her walking between venues in 'Oxbridge', her eating in dining rooms, visiting the library, meeting with a friend, sitting down to write the lecture – all form part of the mind-flow that converged as product in the lecture itself. In *A Room of One's Own*, Woolf described flow moments that were not of Csikszentmihalyian proportions, they did not reach almost orgasmic magnitudes, but they did produce a particular momentum. In focusing on the walks that produced the work, *A Room of One's Own* tracks the inspiration, the surge forward and the completion of a writing project that relied on an account of the creative mind in practice. Woolf's mapping traced the paths and steps she took – physical, intellectual and fictional – and the mind-work she did in writing a classic. Her struggle with the depiction of mind-flow in creating the *A Room of One's Own* lectures reflected the key challenge of her other creative works: her continual attempt to replicate the thinking of characters in emotional and dramatic personal and social situations, as in *Mrs Dalloway* (1925), *The Waves* (1931), and elsewhere.

In 1924, not long before Woolf prepared her Oxbridge lectures in England, Walter Benjamin had an academic dissertation rejected at the University of Frankfurt due to the fact that the mind-flow it foregrounded and analysed was seen by the academy as 'inappropriate' and 'impenetrable', ostensibly due to the 'absence of an uninterrupted purposeful structure' (Steiner, 1998: 11; Watkins & Krauth, 2016). Benjamin protested that while his dissertation did not have a linear structure it was, indeed, more legitimate than a normal submission because it traced the actual working flow of the mind. He admitted that his work had no traditional line of argument, there were gaps in the narrative, the narrative itself kept returning to its starting point, and the whole was a jigsaw of fragments which, in their capricious layout, reflected more the concept of thinking about a topic than the production of a conventional, conclusive thesis (Watkins & Krauth, 2016). But he claimed it was a legitimate dissertation because it wrestled with finding the flow of thought that went into academic thinking. His examiners did not agree. Benjamin had committed the unforgiveable – drawn aside the curtain of respectability which normally hid the messiness of mind-work that lies at the base of great academic thought.

In *One-Way Street* Benjamin took this further. He explored perception, metacognition and recall in the writing process. He traced the movement of rumination, the 'shifting currents' of ideas that appear within the process (Benjamin, 1986 [1932]: 5), the alteration of tiny bits of memory and images as a writer attempts to set them down:

> He who has once begun to open the fan of memory never comes to the end of its segments; no image satisfies him, for he has seen that it can be unfolded, and only in its folds does the truth reside; that image, that taste, that touch for whose sake all this has been unfurled and dissected: and now remembrance advances from small to smallest details, from the smallest to the infinitesimal, while that which it encounters in these microcosms grows even mightier. (Benjamin, 1986: 6)

Benjamin's brilliance lay in his ability to track his mind-flow at its most intimate, and to make writing out of it. Metacognition was the method that Woolf and Benjamin used and recorded. It involved the need to see how one's mind worked in relation to the type of creative product in progress, and the allied activities, such as walking and internal reflection, required to enhance it.

Getting in touch with thought-flow has been the key project for flaneurs and flaneuses since the 19th century, when strolling in cities became associated with writing. Literary criticism has focused attention on the things those writers saw in the passageways of the external world, but creative writing studies can also look at what this tradition tells us about the internal view.

Types of Mind-flow in Writing: Memory, Emotions, Substances

While the dancer and rock climber need their whole body to flow in conjunction with their mind, the scientist, religious mystic and writer might seek flow with less emphasis on the body's cooperation (these are examples mentioned by Csikszentmihalyi). With all forms of work, however, body plays a key role: scientists need their physical senses and actions to operate in coordinated flow with their thinking (experimental data must be properly set up and observed); mystics need their bodies to attain particular positions of flow relationship with their minds (the stance and breathing are essential); and writers confirm the importance of their bodies' contribution in creating the flow for writing (as in the benefits suggested through walking or other physical factors, such as the ergonomics of the writing desk, the substances provided to the body, etc).

And while writing momentum requires our minds to bring us into significant, task-targeted contact with our perceptions, memories and emotions, each of these mind-activities can *interrupt* our creative work, can throw us off the mind-flow we seek. Working at our desks, we can be distracted by the presence of the internet or the view out the window. We can be diverted by the memory that our mother died last year or that our marriage is breaking up. We can be sidetracked by the sadness we feel about how many animals perished in recent wild fires, or the elation we feel about winning a literary award. Additionally, if we resort to substance ingestion to help our minds along, those substances can trip us up as much as provide assistance. Memory, emotions and substance use can be major tools in the writing process, and they can be major hazards.

Memory flow

Creative writers don't 'make it up', as many people popularly believe: writing is not a mind-spew of random, unprecedented concoctions. The thoughts that flow into writing most often come from the gathering and re-combination of issues, facts and events that the writer is familiar with, including those they have purposefully made themself familiar with. While some ideas may pop unexplainedly from the unconscious, writers take the bulk of their writing material from memory of experience and previous thinking... and adapt it. The idea of 'making it up' suggests that the writer sits at their desk letting their mind supply a stream of mind-flow that has no provenance and is the outcome of unconsidered, un-worked serendipity. Nothing is further from the truth about the creative writing process.

Cognitive psychologists talk about various types of memory (long-term, middle-term and short-term, semantic, episodic, autobiographical, proce-dural, sensory, etc.) and how these operate differently in the brain. Each is important for the writing process; we use memory at every stage of writing. We use long-, middle- and short-term memory to recall childhood incidents or last week's events, to remember how we fashioned the last paragraph, or how we just mind-phrased the next sentence we are about to put down. We fashion creative writing from memories swirling at the centre, and the edges, of consciousness, calling on other mind-activity to corral and focus them. When editing, we seek to remember just what we were trying to get our minds around with those earlier-drafted words but, as we revise, we realise we may have changed what we originally wanted to say because we have applied new memories to it. For the content of creative works, we make extensive use of semantic memory (general knowledge), episodic memory (recall of events) and autobiographical memory (closely related to episodic memory but specifically concerned with events in one's personal life).

The harnessing of memory is probably the major skill that writers develop. It is not a skill we need to find new resources for, because in the natural production of healthy autobiographical memory (that aspect of our brain-work where we put together the record of what we have experienced and who we are) we constantly smooth things out, gather diverse bits together, fashion a flowing account to be delivered to others outwardly and to ourselves inwardly, and create a mental narrative that is our personal story. Interruptions and palimpsest occur frequently here: we think about most things from our pasts differently and differently again throughout our lives. Specifically, as creative writers we want memory to provide us with content, analysis and viewpoint. We want memory to supply a flow of useable and pertinent material for the writing project in hand. And we want delivery right now, here, at the desk we type away at.

Particularly significant for mind-flow is *autobiographical memory*:

> that uniquely human form of memory that moves beyond recall of experienced events to integrate perspective, interpretation, and evaluation across self, other, and time to create a personal history. To put it succinctly, autobiographical memory is memory of the self interacting with others in the service of both short-term and long-term goals that define our being and our purpose in the world. (Fivush, 2011: 559)

Autobiographical memory functions to produce the 'coherent narrative' of our life story which eventually tells others, and ourselves, who we are, where we fit, where we've been, and what our accumulated identity comprises. Some mind researchers have objected to autobiographical memory being seen as a 'a quasi-literary work' of the mind, producing a readable 'text' of one's life (Brockmeier, 2015: 175). But it is hard to ignore the fact that the 'mental time travel' the autobiographical memory does for us, the way it engages us with the past, and the 'continuous stream to the present' it produces (Fivush, 2012: 227) are all indispensable aspects of the writing process (Fernyhough, 2014).

With a well-functioning autobiographical memory, we can tackle the task of making meaning flow from the diverse elements of our experience and hopefully combat the divergent interruptions that occur for us when we try to put our past in meaningful order. Plenty of people cannot do this, which leads them to seek therapy. But writers do need that convergent flow capacity that comes from the ability to integrate multiple past events, multiple previous reading and writing experiences, multiple fantasies and emotions and sensations, and generally speaking, the multiplicity of life as it happens to us – not just for the writing of memoir and autobiography but also for novels, poetry, plays, short stories, etc. W.G. Sebald said:

> I think how little we can hold in mind, how everything is constantly lapsing into oblivion with every extinguished life, how the world is, as it were, *draining itself*, in that the history of countless places and objects which themselves have no power or memory is never heard, never described or passed on. (Sebald, 2013 [2001]: 34; my italics)

While Sebald fashioned brilliant fiction based in his memory flow and saw at the same time a general draining away of recall in society, Isabel Allende wrote in her first memoir, *Paula* (1994):

> In the long, silent hours, I am trampled by memories, all happening in one instant, as if my entire life were a single, unfathomable image. The child and girl I was, the woman I am, the old woman I shall be, are all water in the same rushing torrent. My memory is like a Mexican mural in which all times are simultaneous... So it is with my life, a multilayered and ever-changing fresco that only I can decipher... The mind selects, enhances and betrays... (Allende, 1995 [1994]: 23)

Dealing with the tragic circumstances of her daughter's death, the mind-flow became for Allende an overpowering torrent. When emotion and memory combine, the flow conditions are not necessarily conducive to writing. The flow is overwhelming.

Individual differences indicate that memory, like every other function of brain and body, is not the same conscious experience for each of us. Memory will present itself differently to different people. Visual memory, for example, may operate primarily for some as a movie, for others as a series of still shots, for yet others as a montage of both. (How do you remember 9/11? As an image that moves, or as a frozen snapshot? Or both? How do you remember your mother? As an image that moves, or as a frozen snapshot? Or as a voice, a touch, a fragrance perhaps?) I often find my own visual memory, when flowing, takes the form of montage sequences – still shots that progress layer upon layer across one another. That is how I remember my mother. But for 9/11 I see a moving image of a plane approaching and hitting a building, probably because I only ever saw it as film footage. Czeslaw Milosz (1968 [1959]: 3) described how he saw mind-flow: 'I do not see [mind-images] in chronological order as if on a strip of film, but in parallel, colliding with one another, overlapping' – in other words, a complex sort of swirling montage. Often, too, it seems that a variety of memory types lubricates the flow of visual reminiscence:

touch, smell and sound memory (haptic, olfactory and echoic) can enter the mix to make the recollection flow compelling. No matter how a creative writer's memory works, and regardless of what it is they have captured there, memory provides a bank of valuable experience data, a rich source of dramatic material, just waiting to be spliced together in the mind-studio.

Emotional flow

Our emotions and feelings have the power to assist and hinder us, in equal measure, in the writing process. When we are passionate about what we are writing – whether it be due to the message, the form, the characters, the setting, or the aesthetic of the work – the mind-flow generated by emotion can drive us forward. When we are passionate about *something else* than what we are writing, all sorts of mind-flow blocking problems arise. In the midst of the debate about how emotions and the human brain are related, Antonio Damasio (1994) said: 'It does not seem sensible to leave emotions and feelings out of any overall concept of mind. Yet respectable scientific accounts of cognition do precisely that' (Damasio, 1994: 158). Furthermore:

> *feelings are just as cognitive as any other perceptual image*, and just as dependent on cerebral-cortex processing as any other image... I see feelings as having a truly privileged status. They are represented at many neural levels, including the neocortical... But because of their inextricable ties to the body, they come first in development and retain a primacy that subtly pervades our mental life... feelings have a say on how the rest of the brain and cognition go about their business. Their influence is immense. (Damasio, 1994: 159–160; italics in original)

Emotions impact a broad range of human thinking that creative writers include in building and analysing their characters and plots: love, depression, anger, fear, phobia, hate, disgust, grief, sorrow, envy, etc. Experiencing these emotions, recording their effects on the mind, body and behaviour, and being able to describe them authentically, is a mandatory writerly qualification. But experiencing – or re-experiencing – them at the time of trying to write can be difficult for the writer to deal with, and unproductive.

Creativity researchers Fink, Perchtold and Rominger indicate that:

> While there is plenty of research examining links between creativity and specific moods, creativity in an affective context such as creative behaviour in emotion perception or in regulating an ongoing emotional state is largely unexplored. (Fink *et al.*, 2018: 322)

The research suggests that positive effects of emotion on mind-flow produce divergent thinking that enhances creative output. But negative effects of emotion on mind-flow require divergent thinking to solve the problem (Fink *et al.*, 2018: 322). Any writer who has experienced writer's block will agree that the marshalling of a flow of divergent thinking to fix the problem is not an easy matter – whoever's sitting in the control room up there in the brain (certainly not the writer themself, it seems) just won't cooperate. Scientists admit that the area 'still constitutes a major research interest' (Fink *et al.*, 2018: 324) but say that the problem requires cognitive control, which involves 'set-shifting, memory-updating and the inhibition of pre-potent responses' (2018: 322). Yes, but how does one do it?

Creative writers have tried to find solutions, of course. A feature of many books of good advice from successful writers is a section on writer's block. Jon Winokur's *Advice to Writers* (2000) suggests:

> If you are in difficulties with a book, try the element of surprise: attack it at an hour when it isn't expecting it. – H.G. Wells

> When I have trouble writing, I step outside my studio into the garden and pull weeds until my mind clears – I find weeding to be the best therapy there is for writer's block. – Irving Stone

> When I sit down in order to write, sometimes it's there; sometimes it's not. But that doesn't bother me anymore. I tell my students there is such a thing as 'writer's block,' and they should respect it. You shouldn't write through it. It's blocked because it ought to be blocked, because you haven't got it right now. – Toni Morrison

> (Winokur, 2000 [1999]: 190, 192)

These pieces of advice do not specifically explain the problems being experienced by the writer but they do provide strategies for dealing with small-scale mind-flow emotional distractions. One of the most endearing books of advice for overcoming writer's block is Jason Rekulak's *The Writer's Block: 786 Ideas to Jump-start Your Imagination* (2001). Not only is it unusually shaped – more or less an eight-centimetre cube, or 'block' – but also it amusingly recommends so many ideas for getting rid of blockage that the user barely knows where to start. Rekulak himself suggests merely dipping in at random. I did that twice, and I came up with: 'Invent a character whose life is governed by Murphy's Law', and 'Write about the first time you defied your parents.' While Rekulak's book is mainly about inspiration, about getting started on a new project, and precious little about the blockages that occur along the way in a major project, there is nevertheless plenty of writerly information in it. Notably, it starts with a quote from Joseph Heller: 'Every writer

I know has trouble writing' and it ends hundreds of tiny pages later with the writing challenge: 'Create a character who is struggling with writer's block.' To my mind, the best therapy this book provides is an excellent, distracting laugh at the problem of being blocked.

The serious disruptions – the major emotional traumas that interfere with writing flow, and probably interfere with many other aspects of the writer's life – are not to my knowledge well covered in the available literature and may benefit from serious professional therapy. Writer's block is clearly related to the deepest levels of cognition at which creative writers work. Suffering from it, many writers have turned to substances to help them cope.

Substance-induced flow

To achieve enhanced mental flow, creative practitioners have for centuries turned to substances such as drugs and alcohol. Most writers' accounts of ingestion focus on how the benefits are accompanied by dangers. In his prose work, 'The Poem of Hashish' (1860), Baudelaire wrote about hashish intoxication as 'a vast dream' with its special quality being the 'rapid flow of mental images' (Baudelaire, 1951 [1860]: 84). In this drug-fuelled mind-flow state, 'all the forces are in equilibrium so that the imagination, although wondrously powerful, does not drag the moral sense after it...' (1951: 76), so that cultural and religious inhibitions are negated and a more encompassing and authentic view of the world is attained. Baudelaire (1951: 77–78) suggested that such a rush could be achieved by alcohol through 'the solitary and concentrated drunkenness of the writer' or, better still, with hashish and opium, but he admitted that substances created for the writer an 'artificial Ideal' too often accompanied by 'culpable excess'. With some drugs, 'the intelligence, formerly free, becomes a slave... then reason becomes mere flotsam, at the mercy of all currents, and the train of thought is *infinitely more* accelerated' (1951: 107; italics original). At this point of intoxication, the mind-flow becomes an unmanageable flood, not useful to the writer seeking to capture the benefit of enhanced mental agility. Baudelaire also mentioned the after-effects of a hash night: 'The terrible morrow! All the body's organs lax and weary, nerves unstrung, itching desires to weep, the impossibility of applying oneself steadily to any task...' (1951: 118). So, the effect of using hashish was twofold: 'what hashish gives with one hand it takes away with the other' (1951: 122); the inspiration for mind-fluidity in writing can be cancelled by the physical inability to write. Also, for the writer, there is always the danger of long-term addiction, which could be crippling:

'He who has recourse to poison in order to think will soon be unable to think *without* poison. How terrible the lot of a man whose paralyzed imagination cannot function without the aid of hashish or opium' (Baudelaire, 1951: 122; italics original).

I have already written about the effects of alcohol on the writing process in the light of international studies that investigate experimental, biographical and autobiographical data gained from drinker–writers (see Krauth, 2012). I ended that paper with a conclusion titled 'On drinking in a targeted fashion':

> There are times in the writing process which don't matter as much as other times do; there are negotiable times. You know you will come back to [an] idea and change it, you will return to this bit and that bit and edit them. But in the meantime, you want the flow to keep going, you want to produce, you want that outline, that draft, those words on the page, almost anything, it doesn't matter what turns up… you want to see what might be there, you want the investigation to continue. You will come back later and decide whether that session of planning or drafting was worthwhile or not. Alcohol may assist here, experienced writers say, for those who want to use it.
>
> Writers who lost control under alcohol seemingly lost the capacity to bring a sober editing eye to their alcohol-fuelled creativity. They lost the ability to see the difference between the divergent and convergent thinking stages in their writing process. It must be devastating to lose access to the sober you. There needs always to be an operational control room in the alcoholic writer's brain in charge of strategy – in charge of targeting alcohol use towards the parts of the process where it is useful. As one-time drunkard Stephen King said: 'A writer who drinks carefully is probably a better writer… .' (Krauth, 2012)

Alcohol can free the mind-flow into divergent thinking; but it can lead to scattered uncontrol. Carson, in *The Cambridge Handbook of the Neuroscience of Creativity*, says: 'Some experimental studies have found that a dose of alcohol can indeed stimulate certain stages of the creative process, generally during the insight phase' (Carson, 2018: 140). F. Scott Fitzgerald famously elaborated on this idea when he said about his working life: 'First I take a drink. Then the drink takes a drink. Then the drink takes me' (Fitzgerald, quoted in Pritzker, 2012: 392). Pritzker says:

> The number of variables involved leave the relationship between creativity and alcohol with many unanswered questions. Far more work needs to be done to understand the physiological effects of using drugs as well as the psychological dynamic. Innovative methods are needed to research this challenging field. (Pritzker, 2012: 395)

While scientists may be uncertain about the benefits of mind-flow experienced through substance ingestion, personally I think the comment attributed (wrongly) to Hemingway – 'Write drunk, edit sober' – remains pertinent if one is going to use flow-enhancing substances at all. Except I would change it to: 'Write drunk, don't drive, then edit sober.'

The Social Mind

Psychology experiments have studied people when they are supposedly *vacant* – that is, doing and thinking nothing. In describing the stream of consciousness, William James said that we never stop thinking while we are conscious (James, W., 2019: 224–225). Subsequent studies show that there is a network in the brain that fires up during 'vacancy' – the *default network*: 'the set of regions... that turn on when you are doing nothing' (Lieberman, 2013: 18). Lieberman reports that brain scans show we tend to think about *others and their thinking* when our minds are 'at rest':

> the default network directs us to think about other people's minds – their thoughts, feelings, and goals... it promotes understanding and empathy, cooperation, and consideration...
> The repeated return of the brain to this social cognitive mode of engagement is perfectly situated to help us to become experts in the enormously complex realm of social living. (Lieberman, 2013: 19, 21)

Lieberman suggests human evolution required 'the importance of developing and using our social intelligence for the overall success of our species by focusing the brain's free time on it' (2013: 19). Thus, we have a built-in part of the mind that performs social cognitive tasks that involve 'the capacity to think about other people and ourselves' (Lieberman: 2013: 19). The idea of the social mind suggests that our thinking does not stop at an invisible membrane cutting us off from the world but, rather, that it operates with a multitude of thought-links to thought-activities in the world around us and the thinking we know or suspect goes on there. The previously quoted passage (p. 54) from Donna Leon's Commissario Brunetti novel, *A Sea of Troubles* (2002), suggests the degree to which we read literature in order to find out how other people think, and to feed our need to know about it. I suspect we also read nonfiction, memoirs, news, and watch TV, films, social media, etc., all in order to find out how the rest of the world – that is, our world – is thinking.

Cognitive psychologist Vlad Petre Glăveanu (2018), in his study of the creative aspects of the social mind, says: 'the creative mind can only be an "extended" mind... a mind that expands towards its environment'

(Glăveanu, 2018: 300). He develops the idea that the individual creative mind succeeds by being 'extended', 'socialised' and 'embedded' in its society and culture (2018: 305):

> What are the two basic movements of creativity activity? Immersion and detachment. In other words, being fully engaged in action, experiencing it first-hand and becoming absorbed in its sensorial qualities (something Csikszentmihalyi, 1990, referred to as flow) and, at the same time, being able to take a step back, put some distance between self and work, and reflect on it often from a new perspective... The dual-movement framework of creativity is equally psychological, embodied/material, and social. (Glăveanu, 2018: 309)

Glăveanu emphasises the notion of agency:

> To be a creative agent means much more... than to think divergently or to combine information in ways not seen before. It means to be capable of not only appropriating but also *participating* in culture... the shared environment of ideas, practices, and material arrangements in which we are immersed. (Glăveanu, 2018: 313; italics in original)

So, our minds default to thinking about the world and the thinking that goes on there, and as creative writers we are encouraged by it to go out and mingle, perceive, explore, react, take mental notes, because agency is a key part of the work, and it's what our minds are made to do.

Writers involve themselves with the surrounding world because, otherwise, they have nothing to write about. Those who lock themselves away and write purely from their individual mind-work depend anyway upon memories of previous engagement with the world. The idea of the writer stationed at a desk in an ivory tower is mythical: even today, when the world comes to us on the browser, we are severely limited if that is the only engagement we have and the only agency we pursue, because the browser does not (yet) satisfy all our senses. Writers need to read, to engage with ideas broadcast by the various forms of publication; they need to talk with others, to interact with the spectrum of social and cultural thinking; and they need to experience the world first-hand and fully, to connect with all the senses. Even when I am writing a historical fiction, I cannot limit my mind to reading history: I need to know what goes on around me now, to make the work I write relevant to human experience, and to contemporary readership.

The concept of the social mind explains writers out in the world on investigative activities. It explains them reading news websites, journals, history texts, political texts, psychology texts, sociology texts, and the work of other fiction, memoir and poetry writers. It compels them to get out and 'touch and taste and hear and see the world' (Yeats, 1985 [1912]: 107). It says we are already wired to do things this way: it is normal to think about others' thinking, to attempt to explain it, even though it is hard for anyone to know what goes on in another's head. Creative writers have for centuries pursued the book-published thinking of others in order to know how the world works. But, also, the concept of the social mind tells writers to leave the ivory tower, to walk, mingle, adventure, party, observe, chat, interview, note, remember, do and feel – to make creative writing the active agency it needs to be as a record of human physical and intellectual activity.

But do explorations with our social mind lead us to convergent or divergent thinking? And what are the implications for our writing in the moment when we walk or travel? I think one of the reasons why walking works as an adjunct to writing is that, in walking, we extend ourselves into the reach of our default thinking and we find ideas that correspond with our normal thinking. This is writerly 'soft walking', if you like: extending ourselves into comfortable and supportive environments where we can examine what our minds do 'at rest'. Rousseau's rural walks promoted reveries, calm reflections on the world, as did Wordsworth's and Thoreau's. But writers might also do 'hard walking': walk out in war zones, bushfires, hurricanes, volcanic eruptions, earthquakes and tsunamis… and cities. As a rule, writers don't do it *en masse*: they do it singly, as agents keen to experience particular environments for specific writerly purposes. (One of the best things I ever did was walk out during a cyclone and use that thinking in a story I wrote later.) The early Romantics walked out and found their thinking about nature at a time when industry was ramping up alarmingly; *flaneurs* and city walkers went out and contacted new cultural thinking at a time when international political awareness was being transformed in disturbing ways. Whether the walking is soft or hard, the writer's aim is to get in touch with the world-mind, to read the world of experience and thinking, and to analyse it.

So, does exploring the world-mind produce flow-inspiring convergent or divergent thinking? Clearly both. Some experiences, in contemplation, make sense and line up in cogent argument. This is the easier stuff. Others do not, and need further thinking, further manipulation, which is the domain writers work in. Albert Camus, in a 1957 speech made after being awarded the Nobel Prize in Literature, said: 'In the midst of such din [i.e. what's going on in the world] the writer cannot hope to remain aloof in order to pursue the reflections and images that are dear to him'

(Camus, 2018 [1957]: 1). Camus stressed the writer's obligation to engage their mind with the surrounding political, philosophical and social worlds. He called the process 'creating dangerously', and concluded his speech with:

> Great ideas, it has been said, come into the world as gently as doves. Perhaps then, if we listen attentively, we shall hear amid the uproar of empires and nations, a faint flutter of wings, the gentle stirring of life and hope. Some will say that this hope lies in a nation; others in a man [sic]. I believe rather that it is awakened, revived, nourished by millions of solitary individuals whose deeds and works every day negate frontiers and the crudest implications of history. (Camus, 2018: 32–33)

Camus' speech about the obligations the social mind imposes on writers allows us to see how useful this aspect of our thinking is for writing. We are part of a world of thinking; we all have the capacity to tap into it; it provides a plethora of matter highly useful for writing; it represents a massively challenging extension of ourselves while also it is a moment away from our personal thinking.

Our social mind is a key factor in our agency as writers. Our writerly minds may flow with relative ease through memory of personal experiences, or we may get focus by physical processes such as walking, or indeed we might judiciously enhance mind-flow with the addition of mind-altering substances. But the key point is that, as writers, we need to develop a relationship with our minds so that we engage with more than the limited movement implied in us staying at a work desk in a comfortable environment only looking inward.

Coda

I completed this journey into my mind with the help of many others who undertook similar exploration into their, or others', minds. Two books I looked at on this engaging trip were Sue Woolfe's *The Mystery of the Cleaning Lady* (2007) and Alice W. Flaherty's *The Midnight Disease* (2005a). Woolfe said she was prompted to write her exegetical book because she 'became so stranded' by writer's block (Woolfe, 2007: 2); Flaherty said her book – which is entirely devoted to writer's block – was written to 'show that science can help us create and understand literature' (2005a: 265). As the previous chapters show, I was more inspired as a creative writer/researcher by novelist Woolfe than by neurologist Flaherty, in spite of the latter's excellent reviews on publication. For me it was, ultimately, a matter of shared language, experience and world view. Creative writer Woolfe engaged my mind more deeply than did medically-oriented

Flaherty. I was prompted to write *this* book from a desire to know what goes on in the creative writer's mind not only when blocked but when things are going swimmingly, when some sort of flow is happening, and impart my findings to other creative writers. I am interested in what my mind does when I write, because I want to learn as much as I can about my writing practice. I have in the past attempted to 'observe', or metathink, by analysing my own thinking, and I'm sure this is not unusual for writers keen to further their skills. As a teacher of creative writing, it would be wonderful, one day, to be able to tell my students *exactly* what happens in their brains when they write, and how that differs from what happens for expert writers. Or how it will happen *to them* when they become expert.

So, I arrive now to the end of this work more interested in writer's block than I ever was because, like Sue Woolfe, I still don't see clearly dependable ways out of it; and a highly talented Masters student of mine has just bombed out because all he could ever get onto the screen were brilliantly elaborated project proposals. When it came to producing the real thing, his capacities dissolved. So, I think it appropriate to say a final, personal few words about writer's block.

How we each deal with writer's block is based in our personal capabilities and circumstances. For each there is an individual, tailored solution. But who will tailor it? Most likely, we must do it for ourselves. And to succeed in this, we must become extraordinarily aware and insightful about how our own mind works. I have noted each day I spent writing this book that I had difficulty getting going. I figured out, via spotting things I would do rather than writing, that (a) I worked best straight after waking up in the morning, when my mind was fresh, and could then keep going for hours; and (b) I could especially keep the engagement going successfully provided I did not turn the television on during any brief break I took.

I realise this is a highly personal analysis of the writer's block problem – and that is my point. Ideas contained in this book converge at this point. If we know more about our own minds' operation during writing, then we can analyse our own processes and be better writers. I intend to research further the scientific view on (a) writing creatively in the morning rather than in the afternoon because of the brain waves produced in that part of the day, and (b) the kind of brain waves produced by television, because I can feel them doing my writing brain waves no good. Of course, next week or next year my temporary blocks might be produced by something else entirely.

Finally, I have to say, I enjoy every successful day planning, drafting and editing in that amped up writing-production studio in my head. Frankly, I love being there.

6 Reflective Questions for Developing Writers and Classroom Discussions

These questions are allied to the research discussion in the chapters of this book. They can provoke mind-exercises, practical writing exercises, or conversations. They are aimed at writers, teachers of creative writing, and classes of writing students.

I hasten to point out that whenever I ask a creative writing class of mine to complete a written exercise, I do it myself alongside them. I also read out my response in turn when they read theirs – because I always learn so much about what I am asking them to do from doing it with them. Then, too, my ego is involved: I want to model for them an impressive response to the exercise. Of course, I regularly find that what they produce is equal to, or better than, mine.

Introduction

The writer's mind: What happens when writing happens? (*Pages 1–3*)

- We talk about the creative writing process but it is a multiplicity of processes. What processes are you aware of going on in your mind when you write? In which situations do they vary? And how do they vary?
- Would you like to know what your brain does when you write? How do you think that might help – or hinder – you?

Creativity and the writer's process (*Pages 3–4*)

- Do you enjoy the kind of thinking that happens when you get your mind to link, for example, the word *table* with the word *ocean*? How might this kind of thinking help you to be a better creative writer?

Recent science about creative writing processes (*Pages 4–8*)

- When you write, does it feel as if a particular part of your brain is doing the work? How does this seem to fit with what the science is saying?
- What do you see as the major benefits of science studying the writer's mind? And what are the major challenges, in your view? Are these benefits equal for the creative practitioner and for the creative writing researcher?
- Who would you trust more to tell you about your writing mind – a creative writer, a scientist, a creative writer–scientist, or someone else entirely? Why?

Research by creative writers (*Pages 8–10*)

- If you were to write specifically about your mind at work, how would you do it – in nonfiction (e.g. an essay) or in creative writing (e.g. a story, a poem, a playscript)? Why? Or might a blended-genre work be best? Why?

Metacognition: Writers thinking about thinking (*Pages 10–13*)

- Rousseau described his writing mind as a chaos which became, with work, much more ordered. He used the metaphor of the Italian Opera stage. What metaphor would you find most appropriate to describe your writing mind?
- Do you use metacognition as part of any activity in your life? If you do it for writing, what parts of the writing process does it tend to focus on? Do you think meta-thinking can help you improve your writing? In what ways?

1 Depictions of the Creative Writing Mind

Muses (*Pages 18–22*)

- When you write, are you thinking that what you are doing is inspired by a divine muse? Does this belief assist you in your writing? How do you prepare for and operate within that paradigm? If you were told that gods don't exist and this muse-notion is actually something else happening, to what would you attribute your 'inspired' feelings?
- The ancient idea of muse-powered writing involved the idea of madness. Must one be mad – or at least just a little mad – to be a

creative writer? Do you know any writers who are mad? How would you describe their madness?

- How much of writing is impassioned transcendence, and how much is sheer hard work?

- Writing can be done spontaneously, allowing the mind to spew forth uninhibited. How valuable is this kind of writing? What is its place in a writer's array of capabilities? Do you agree with the idea that a writer should be able to write differently in different circumstances and for a range of different outcomes? Why?

- What are the outcomes of having a human muse? In what ways is the idea of being inspired by a human muse different from having a divine one?

- Exercises:

  - Invoke the divine Muse and see what happens! Write a poem as given to you by the operation of the Muse in your mind.

  - Decide someone is your muse and see what happens – start writing and keep writing by invoking that person/muse.

  - Decide that a feature in the environment is your muse and see what happens by getting intimate with it and letting it speak. Write what your muse has said to you (i.e. when you let a tree, a wall, a tractor, a skyscraper, etc., speak itself in your mind)?

  - Go out and embrace a muse, whether it be person, statue, car, vegetable, landscape, or a piece of music. Write what you get from your muse through *all* your sense receptors.

  - How would you describe in your own personal language the effects your muse/s had in your mind in the above exercises? And how would you describe in more scientific or academic language (any attempt will do) the effects your muse/s had in your mind in the above exercises?

Sciences (*Pages 22–32*)

- How would you create a diagram to represent your personal writing process? What would be the major stages in your version of a Wallas-like list for writing? How would you develop each of those stages to include activities within each stage? Does your final diagram represent how you always write? If not, how can you adapt it to provide a comprehensive analysis of your process possibilities? Do you think your diagram is useful to you in understanding your writing practice? In what ways is your diagram a portrait of your mind? Is your diagram in the form of a wheel? If not, why not?

- Do you agree that scientific study (or any study) of the mind can help a writer develop their writing process? Or is it safer not to question what happens, for fear of demystifying the process?
- Do you believe that your psychological profile defines aspects of your writing process? Has your personality, childhood, ability to communicate, or any past trauma, affected how, what or when you write?

The literary imagination (*Pages 32–34*)

- If you deconstruct your imagination or imaginative processes by use of metacognition, how would you describe what goes on in your mind when you imagine? Then, taking this further, how do you describe what happens when you use your imagination in the process of writing?

Contemporary depictions (*Pages 34–41*)

- When you write, what parts of you do you feel are involved in the process? A small part of your mind or brain? A great deal of your mind and brain? More than your mind and brain, including, say, parts of your body? More than your mind, brain, and body?

2 Writers and Thinking, According to Critics

The complexity of the stream of consciousness (*Pages 43–48*)

- Popular how-to-write books suggest that for building a character the budding author should make a list that includes such things as the colour of the character's hair and eyes, their body type, their star sign, their favourite food, drink, dog and sport, their least favourite food, drink, dog and sport, and so on. Using such a list, what sort of character do you think the novice writer will produce? Are such lists useful for any genres in particular? Might such lists actually be about how the potential character *thinks*? Would a useful way of shortcircuiting the process be to say to the budding writer: make a list of your prospective character's *thoughts*?
- How should a writer represent thoughts in the head – as a streaming or as a series of separations? Why?
- How best to describe a character's consciousness – with cliched thinking or with thinking unique to the character? What kind of thinking does the genre you write in expect you to use? How much

should you foreground a character's psychology as opposed to pushing it into the narrative background? Are your characters' minds essential parts of the drama in your work?

- In dealing with a specific character's head, should a writer operate metacognitively, using their own understanding of thinking, or use cogent and smoothed-out logics provided from elsewhere – from cultural creed, religious dogma, folk psychology, political suasion, and so on?
- If you were to write down exactly what your own mind's stream of consciousness spewed forth, what would you end up with? Would the result of that exercise be useful knowledge for your creative writing practice when examined?
- Pick a topic – anything – and write a paragraph about it using convergent thinking. Then write a paragraph on the exact same topic using divergent thinking. Follow that with a paragraph on a different topic in either convergent or divergent form, then *edit it* to the other form. What do you notice your mind doing in these exercises?
- With students, I use an exercise I call 'The 3-card trick', based in the idea that any three settings placed in order will produce a plot and characters for a dramatic story. I ask the class for three unrelated settings, for example: (1) the back seat of a car; (2) a home kitchen; (3) an abattoir. I then ask them to outline a story prompted by the three settings. Following that, I ask them to reverse the sequence of settings and come up with the new prompted story, for example based on: (1) an abattoir; (2) a home kitchen; (3) the back seat of a car. Following that, I ask them to suggest replacements for the final setting to produce (a) a happy ending or (b) a tragedy. The exercise provokes the mind to divergent thinking outcomes (the settings) and then to convergent thinking links between them (the plot). Inevitably, narratives are created by the cultural coding available in any given setting. I also ask students to convergently change the narrative by purposeful insertion of a particular setting: for example, what if a graveyard, or a hayloft, or a parliamentarian's office, is introduced at (1), (2) or (3)?

Critics on creative writing about thinking (*Pages 48–58*)

- In your opinion, what are the major issues and qualities of an author's depiction of a character's mind? As a reader what are you most convinced by? As a writer what are you drawn to as authentic ways to go about it?
- Do you consider that your mind works in ways similar to other people's minds? Whether yes or no, what evidence do you have to support your

view? If you see there are variations between minds, what are the reasons for these?

- Sartre called the mental editing we do when writing 'doctoring'. He wasn't talking about the normal process of editing we do once words are on the page, he was talking about what we do to words and ideas in our minds before they even get drafted. Are you aware that this is what you do? What principles, strategies, techniques do you apply in doing this 'doctoring' when writing creatively? Are they different when you write in non-creative forms?

The mind's eye, the inner voice, and other mind processes producing writing (*Pages 58–64*)

- How much of your creative writing involves ekphrasis, i.e. the translation of a nonverbal sense mode into the verbal? And how much do you suppose the expanded notion of the ekphrastic process might be involved for visual artists, music composers, filmmakers, etc. Can an artist paint or sculpt a verbal idea, a piece of music, a touch? Can a composer create music to represent a landscape, a poem, a fragrance?
- How useful to creative writing is the idea that we think dialogically? If you find that the voices in your head are often in disagreement – *Will I eat this chocolate? No, you shouldn't* – might this phenomenon give helpful insight into character creation, plot dramas, and the overall ethical message of a work? Or is this aspect of our mind-activity only useful in writing essays where we say: *This is the argument I want to make; these are the arguments against it*?
- As did ancient notions of ekphrasis, contemporary notions acknowledge that ekphrastic translation can occur between any of our sense modes. Do you find that you are drawn to any of these possibilities in particular? For example, do you particularly enjoy turning touch into words? Or turning touch into visuals into words? Or turning music into words? And so on.

3 Thinking and Writing, According to Writers

Writers' representation of verbal thinking (*Pages 67–78*)

- Creative writers regularly depict the thinking of their characters, and in order to do so they must examine their own thinking. But not all characters are given minds that work exactly as their creators' minds do. How does the creative writer 'perform' the thinking of an 'other' in their own mind? What mind-shifts must the author accomplish to represent the thinking of a character unlike themselves?

- When writing, do you consider the thinking of your reader? How does that influence what you write? What do you depend on the reader for? To what extent do you write to accommodate the reader? Is that different when writing in different genres?

- Representing the verbalised stream of consciousness of a character is significantly a matter of capturing a voice. Do you think capturing the speaking voice of another will help to create the thinking voice of a character based on that person? Do you think you think with a voice similar to the voice you speak out with?

- Which is the most effective way to represent the mind-voice – as a logical sequence of thoughts, or as a disrupted, fragmented progression? What genre requirements impact the idea of 'effectiveness' here?

- Can a male author write the thinking of a female mind? And vice versa. Can an author from an older generation write the thinking of a younger-generation mind? And vice versa. Can a person of one race or culture write the thinking of a person of another race or culture? How do your answers to these questions impinge upon the effectiveness and significance of the creative writing art form as a whole?

- It can be suggested that creative writing provides a release of thoughts from the darkest corners of writers' minds. Is this a good or bad thing? And on what do you base your notion of 'good' and 'bad'?

- In poetry or prose written in concrete/pattern form (à la Federman), write a page-long work depicting the thoughts in your mind as you write the piece.

Writers' representation of visual thinking (*Pages 78–90*)

- Consider the idea of a story or poem which starts with the following as a mind-experience: (1) a single image; (2) a single word or sentence; (3) a single touch; (4) a single smell; (5) a single taste; (6) a single sound; (7) a single emotion; (8) a single memory. Can each of these radiate out into a full story or poem? If so, by what processes does this happen?

- How *whole* should the idea for a story or poem be, before an author starts drafting it?

- Can a succession of mind-images emanating from an initial visual image indicate the narrative structure, language choice and rhythms, and character development for a whole written work?

- If you experience visual images in your mind, where do you think they mostly come from – your lived seeing, your memories (probably altered over time), the memories of others told to you, your fantasies, your dreams, photographs you have seen, images available in the cultural understanding around you?
- French writers speak of *paysage intérieur* – the idea that we each have a personal landscape in our minds. So, for some of us (who always think negatively) our generalised internal landscape might be a barren wasteland; for others (who think very positively) it is an enduring paradise. In developing a character for a creative work, how might this idea about the visualised country within help with developing the character's mind?
- Exercises:
  - Bring up on your mind-screen a scene that involves action, for example: domestic violence in a bedroom; or a young couple in the back seat of a car. Write a poem or story about the scene *in slo-mo*. Note: the action that occurs – the visualising of it in your mind – has to happen in slow motion as you write it down.
  - Use the same scene you used in the previous exercise and describe the action by significant use of metaphors. These metaphors should translate between the senses for plausible reasons related to the characters' thinking.
  - Writers don't use setting merely to locate a story or poem geographically: setting is coded with additional meaning. (For example, if the setting for a story is a high-rise building in Florida, what is the impact of calling the building 'The Parthenon'?) Write a short scene where the setting is key to understanding what is going on.
  - Write a piece where the main characters do not appear but their personalities and propensities are created by description of the place where they live. The narrative will take the viewpoint of someone who has broken into their abode while they are out.
  - Bite into a fruit or vegetable and write a haiku or six-word story. Then bite into a piece of pork or beef and write a haiku or six-word story.
  - Produce a story or poem that involves the drama and memorability of a coloured smell or an audible taste.

Writers' representation of other sense modes (*Pages 90–98*)

- In order to fully describe the touch, taste, or smell of an experience, should the writer leave their desk and take a notebook to go out, find and describe the experience *in situ*? To represent sensory experience, which is best – the mind-image created by spontaneous and immediate experience, or the mind-image created by memory?

- Pseudo-synaesthesia – where a writer describes one sense mode in terms of another via managed processes in the mind – is of interest to those who want to create distinctive, impactful metaphors. These might include, for example, the colour of a sound or the texture of a smell. But how does this managing process happen in the mind? What metacognition is involved for the writer making metaphor in this way?
- Each of the senses comes to us with different qualities and associations. Does the language we need to describe them – and the mind-work of ekphrasis – require that we treat each sense differently?

4 The Mosaic Mind: Writing and Divergent Thinking

The mosaic mind (*Pages 99–106*)

- In other cultures, the gap, the blank, and spaces between statements can be read positively. Do you think western, English-speaking culture can develop this type of perception? How will they do it?
- When editing a creative piece, do you ever consider the idea of editing out a significant aspect of the narrative? Under what circumstances might you do this? What mind-work do you thereby place on the reader?
- In the plotting phase for a work, brainstorming and planning require significant divergent thinking. But convergent thinking – bringing all the possibilities and consequences together – is also required. How do you manage this tightrope balancing exercise?

'Things fall apart': The rise of mosaicked writing (*Pages 106–115*)

- Tristan Tzara's cut-up method for writing a poem tells us about how our minds work to make meaning. As with the three-card trick mentioned above, when our minds are confronted with adjacency, they use divergent Jamesian 'fringes and haloes of relations' to create thinkable links. Are you aware of your mind working in this way? Create a random cut-up poem and monitor (internally) as your mind reads it.

'Minding the gap': The mosaic work of writers (*Pages 115–120*)

- When you leave a gap in a narrative, how much do you think through the consequences of what you are doing? What do you rely on to navigate the writing spaces in a work, and how do these relate to what you think readers will do?

- Are you the kind of creative writer who is happy to invite multiple conclusions to the reading of your work? If so, from where do you get the confidence to do so? If not, explain your doubts. How do you understand the reading process happens?

The wandering mind: Writing and fantasising (*Pages 120–123*)

- Does mind-wandering assist or hinder you? If it assists, are there specific circumstances in which this is so? If it hinders, when so? Do you think there may be ways to divert mind-hindering wandering so that it becomes *mind-assisting* wandering? For example, when you are feeling mind-hindered, might writing about your metacognitive thinking provide a way out?
- Writer's block is a serious function of cognition interference. Considering the several approaches this book has taken regarding metacognition, the nature of the stream of consciousness, and the dynamics between divergent and convergent thinking, do you think writer's block might be theorised, thought through, and conquered? Why?

5 The Flow Mind: Writing and Convergent Thinking

The flow mind (*Pages 124–129*)

- Creative writers who achieve Csikszentmihalyi's (1996) flow experience, have likened it to having a writing orgasm. Has this ever happened to you? If so, how might *you* describe the feeling of it? Recall the best day of writing you ever had and give the reasons why.
- What have been the conditions in which you found yourself most able to write successfully? Were they related to your environment, to the nature of the project, to personal circumstances, to your emotional state, to substances ingested, etc? How do these conditions relate to Csikszentmihalyi's (1996) list of elements?
- Any of Csikszentmihalyi's list of elements might form the basis for a discussion amongst creative writers, and for a writing exercise to test its usefulness. I will choose just the three questions I posed previously:
  - (1) When does the creative writer ever know exactly where they are going?
  - (2) How do we ever know how to treat feedback?
  - (3) How do we ever know we are up to the task of assessing our own capabilities?
- Do you ever think that fashioning a creative work is a problem-solving exercise? In what circumstances does this occur?

- What difference is discernible in the way you think in the flow of writing creatively as opposed to the strategies you bring to editing that work?

Writing, walking and the convergent mind (*Pages 129–137*)

- Does walking work for you as a way to boost mind-flow in the writing process? In which stages of the process do you find walking most helpful? Do you think differently when you walk? How would you describe the effect on your mind that walking has? What type of place or scenery do you prefer to walk in to kick-start, augment, continue, or improve your writing? Why do you think that particular environment assists you? If walking doesn't do it for you, what does?

Types of mind-flow in writing: Memory, emotions, substances (*Pages 137–145*)

- How important is memory to you in your writing process? If you use it as a major source of inspiration, do you write only about yourself, e.g. in memoir form? Or do you apply your own memories to other characters in a fictional way? And what mental processes do you use to *dredge* your memory? Do the memories you want pop up easily? Or do you strategise going in search of them? How do you do that?
- When you have your next poem, story or play in mind – cooking away there, as it were, in memory – do you tell others about it? Or do you keep it secret to yourself because to set it free from writerly memory means you have lost it? Why?
- Do you find that you write better in certain emotional states as opposed to others? If so, how might you explain this in terms of what goes on in your mind? Have you ever found a way of dealing with negative or disruptive feelings so that writing can proceed? Can this be explained, do you think, in terms of divergent or convergent thinking?

The social mind (*Pages 145–148*)

- When you are 'thinking about nothing', what do you think about? Does your mind turn to others around you and what they are thinking, as the research suggests?
- Clearly, creative writers operate in the context of, and in response to, what goes on in the world. (For example, different genres are targeted at readers in different parts of society.) But to what extent are you conscious of the fact that every word you write came into your mind from the outer world, and goes back out there for reception? Or do some words, indeed, come from you entirely?

References

Abraham, A. (2018) *The Neuroscience of Creativity*. Cambridge: Cambridge University Press.

Abrams, M.H. (1971) *The Mirror and the Lamp: Romantic Theory and the Critical Tradition*. Oxford: Oxford University Press.

Ackroyd, P. (1998 [1986]) A killer, haunted by smells. *The New York Times on the Web*: https://movies2.nytimes.com/books/97/04/06/reviews/ackroyd-suskind.html (accessed 24 August 2020).

Adorno, T. (1951) *Minima Moralia: Reflections from a Damaged Life* (E.F.N. Jephcott, trans.). London: Verso.

Alexievich, S. (2017 [1985]) *The Unwomanly Face of War*. London: Penguin.

Allende, I. (1995 [1994]) *Paula* (M.S. Peden, trans.). New York: HarperCollins.

Allende, I. (2013) Chapter one: Isabel Allende. In M. Maran (ed.) *Why We Write: 20 Acclaimed Authors on How and Why They Do What They Do* (pp. 4–11). New York: Plume.

Amabile, T.M. (2019) The art of (creative) thought: Graham Wallas on the creative process. In V.P. Glăveanu (ed.) *The Creativity Reader* (pp.15–32). New York: Oxford University Press.

Armstrong, P.B. (2013) *How Literature Plays with the Brain: The Neuroscience of Reading and Art*. Baltimore, MD: Johns Hopkins University Press.

Armstrong, P.B. (2019) Neuroscience, narrative, and narratology. *Poetics Today* 40 (3), 395–428. DOI: 10.1215/03335372-7558052.

Barbot, B., Tan, M., Randi, J., Santa-Donato, G. and Grigorenko, E. (2012) Essential skills for creative writing: Integrating multiple domain-specific perspectives. *Thinking Skills and Creativity* 7 (3), 209–223. https://fdocuments.net/document/essential-skills-for-creative-writing-integrating-multiple-domain-specific.html (accessed 20 February 2021).

Baron-Cohen, S. and Harrison, J. (1997) Synaesthesia: An introduction. In S. Baron-Cohen and J. Harrison (eds) *Synaesthesia: Classic and Contemporary Readings* (pp. 3–16). Oxford: Blackwell.

Barthes, R. (1978 [1977]) *A Lover's Discourse: Fragments* (R. Howard, trans.). New York: Hill & Wang.

Baudelaire, C. (1951) The Poem of Hashish [1860]. In P. Quennell (ed.) *My Heart Laid Bare and Other Prose Writings* (pp. 73–123; N. Cameron, trans.). New York: Vanguard Press.

Baudelaire, C. (1993 [1857]) Correspondences. In Charles Baudelaire, *The Flowers of Evil* (p. 19; J. McGowan, trans.). Oxford: Oxford University Press.

Baudrillard, J. (1997) *Fragments: Cool Memories III, 1990-1995* (E. Agar, trans.). London: Verso.

Bayer, M. (2010) Q & A. *Berlin School of Mind and Brain Newsletter No. II: An Issue on Language* (June), 5–17. http://www.mind-and-brain.de/fileadmin/downloads/Ausschreibungen/mab_newsletter_no_02_nachdruck_2_2011_screen.pdf (accessed 5 August 2019).

Beaty, R.E., Benedek, M., Silva, P.J. and Schacter, D.L. (2016) Creative cognition and brain network dynamics. *Trends in Cognitive Sciences* 20 (2), 87–95.

Benedek, M. (2018) Internally directed attention in creative cognition. In R.E. Jung and O. Vartanian (eds) *The Cambridge Handbook of the Neuroscience of Creativity* (pp. 180–194). Cambridge: Cambridge University Press.

Benjamin, W. (1986 [1932]) A Berlin Chronicle. In P. Demetz (ed.) *Reflections: Essays, Aphorisms, Autobiographical Writings* (pp. 3–60). New York: Schocken Books.

Benjamin, W. (1998 [1928]) *The Origin of German Tragic Drama* (J. Osbourne, trans.). London: Verso. http://rebels-library.org/files/benjamin_drama.pdf (accessed 20 November 2015).

Benjamin, W. (2002) *The Arcades Project* (H. Eiland and K. McLaughlin, trans.). Cambridge, MA: The Belknap Press of Harvard University.

Benjamin, W. (2016 [1928]) *One-Way Street* (E. Jephcott, trans.). Cambridge, MA: The Belknap Press of Harvard University.

Bilman, E. (2013) *Modern Ekphrasis*. Bern: Peter Lang.

Bouchet M. (2020) Neurological synesthesia vs literary synesthesia: Can Nabokov help bridge the gap? In M. Bouchet, J. Loison-Charles and I. Poulin (eds) *The Five Senses in Nabokov's Works* (pp. 255–274). Cham: Palgrave Macmillan. https://doi.org/10.1007/978-3-030-45406-7_16.

Bowling, L.E. (1950) What is the stream of consciousness technique? *PMLA* 65 (4), 333–345. https://doi.org/10.2307/459641.

Bradbury, R. (1996) *Zen in the Art of Writing*. Santa Barbara, CA: Joshua Odell Editions.

Brady, T. (2000) A question of genre: De-mystifying the exegesis. *TEXT: Journal of Writing and Writing Courses* 4 (1): http://www.textjournal.com.au/april00/brady.htm (accessed 10 August 2020).

Breedlove, J.L., Ghislain, S.Y., Olman, C.A. and Naselaris, T. (2020) Generative feedback explains distinct brain activity codes for seen and mental images. *Current Biology* 30 (12), 2211–2224. https://doi.org/10.1016/j.cub.2020.04.014.

Breton, A. and Soupault, P. (1997 [1920]) *The Magnetic Fields*. In A. Breton, P. Eluard and P. Soupault, *The Automatic Message/The Magnetic Fields/The Immaculate Conception* (pp. 55–145) (D. Gasgoyne, A. Melville and J. Graham, trans.). London: Atlas Press.

Brockmeier, J. (2015) *Beyond the Archive: Memory, Narrative and the Autobiographical Process*. Oxford: Oxford University Press.

Brophy, K. (1998) *Creativity: Psychoanalysis, Surrealism and Creative Writing*. Carlton South, Victoria: Melbourne University Press.

Brophy, K. (2009) *Patterns of Creativity: Investigations into the Sources and Methods of Creativity*. Amsterdam: Rodopi.

Bruhn, S. (2000) *Musical Ekphrasis: Composers Responding to Poetry and Painting*. Hillsdale: Pendragon Press.

Brunner, J. (2015) Famous writers share how they handle writer's block. *HuffPost* (31 January): https://www.huffpost.com/entry/famous-writers-share-how-_b_6560188 (accessed 18 November 2019).

Burroughs, W.S. (2012) *The Western Lands*. London: Penguin.

Burroway, J. (2011 [2003]) *Writing Fiction: A Guide to Narrative Craft. Eighth Edition* (with E. Stuckey-French and N. Stuckey-French). Boston: Longman.

Calvino, I. (2016 [1988]) Visibility. In I. Calvino *Six Memos for the Next Millennium* (pp. 99–121) (G. Brock, trans.). London: Penguin.

Cameron, S. (1989) *Thinking in Henry James*. Chicago. IL: University of Chicago Press.

Camus, A. (2018 [1957]) Create dangerously. In A. Camus, *Create Dangerously* (pp. 1–33) (Justin O'Brien trans.). London: Penguin. (Originally a speech delivered at the University of Uppsala, Sweden, December 1957.)

Carson, S.H. (2018) Creativity and psychopathology: A relationship of shared neurocognitive vulnerabilities. In R.E. Jung and O. Vartanian (eds) *The Cambridge Handbook of the Neuroscience of Creativity* (pp. 136–157). Cambridge: Cambridge University Press.

Cheeke, S. (2008) *Writing for Art: The Aesthetics of Ekphrasis*. Manchester: Manchester University Press.

Cohn, D. (1978) *Transparent Minds: Narrative Modes for Presenting Consciousness in Fiction*. Princeton, NJ: Princeton University Press.

Coleman, P. (2008) Introduction. In J-J. Rousseau *Confessions (Les confessions)* (A. Scholar, trans., P. Coleman, ed., pp. vii–xxix). Oxford: Oxford University Press.

Coover, R. (1969) The babysitter. In R. Coover *Pricksongs & Descants* (pp. 206–239). New York: Grove Press.

Coover, R. (2005) Heart suit. *McSweeney's* Issue 16.

Corballis, M.C. (2018) Laterality and creativity: A false trail? In R.E. Jung and O. Vartanian (eds) *The Cambridge Handbook of the Neuroscience of Creativity* (pp. 50–57). Cambridge: Cambridge University Press.

Cropley, A.J. (2011) Teaching creativity. In M.A. Runco and S.R. Pritzker (eds) *Encyclopedia of Creativity. Volume 2* (2nd edn, pp. 435–445). Amsterdam: Academic Press.

Csikszentmihalyi, M. (1996) *Creativity: Flow and the Psychology of Discovery and Invention*. New York: HarperPerennial.

Damasio, A.R. (1994) *Descartes' Error: Emotion, Reason and the Human Brain*. New York: Putnam.

D'Annunzio, G. (2011 [1921]) *Notturno* (S. Sartarelli, trans.). New Haven, CT: Yale University Press.

Day, A. (1998) Ma: The Japanese spatial expression. *The Golden Century, 1562-1657*: http://www.columbia.edu/itc/ealac/V3613/ma/ (accessed 19 January 2020).

Denham, R.D. (2010) *Poets on Paintings: A Bibliography*. Jefferson, NC: McFarland & Co.

Didion, J. (1976) Why I write. *New York Times Book Review* (5 December), 98–99. Available at: *qwriting*, theessayexperiencefall2013.qwriting.qc.cuny.edu/.../"Why-I-Write"-by-Joan-Didion (accessed 28 December 2016).

Doctorow, E.L. (1986) The art of fiction no. 94. Interview by George Plimpton. *The Paris Review* 101 (Winter): http://www.theparisreview.org/interviews/2718/e-l-doctorow-the-art-of-fiction-no-94-e-l-doctorow (accessed 29 December 2016).

Dodd, A. (n.d.) Muse: The brain sensing headband. *Anxiety Attack Solutions*. https://anxietyattack.solutions/muse-the-brain-sensing-headband/ (accessed 28 January 2021).

Donne, J. (1896 [c. 1590]) The perfume. In E.K. Chambers (ed.) *The Poems of John Donne*. London: Lawrence and Bullen. Available at: *Bartleby.com*: https://www.bartleby.com/357/65.html (accessed 8 February 2021).

Donnelly, D. (2018) The convergence of creative writing processes and their neurological mapping. In G. Harper (ed.) *Changing Creative Writing in America: Strengths, Weaknesses, Possibilities* (pp. 95–111). Bristol: Multilingual Matters.

Duffy, P.L. (2020) Colorful language: How synesthetes perceive words. *Brain World* (26 May): https://brainworldmagazine.com/colorful-language-how-synesthetes-perceive-words/ (accessed 14 August 2020).

Dujardin, E. (1990 [1887]) *We'll to the Woods No More* (S. Gilbert, trans., 1938). New York: New Directions.

DuPlessis R.B. (1990) *The Pink Guitar: Writing as Feminist Practice*. New York: Routledge.

Edel, L. (1964 [1955]) *The Modern Psychological Novel*. New York: Grosset & Dunlap.

Edel, L. (1982) *Stuff of Sleep and Dreams: Experiments in Literary Psychology*. London: Chatto & Windus.

Edel, L. (1990 [1957]) Introduction. In E. Dujardin *We'll to the Woods No More* (pp. vii–xxvii). New York: New Directions.

Eidt, L.M.S. (2008) *Writing and Filming the Painting: Ekphrasis in Literature and Film*. Amsterdam: Rodopi.

Enck, J.J. and Hawkes, J. (1965) John Hawkes: An interview. *Wisconsin Studies in Contemporary Literature* 6 (2), 141–155. DOI: 10.2307/1207254

Epictetus (1948 [c. 135 AD]) *The Enchiridion [The Handbook]* (T.W. Higginson, trans.). New York: Liberal Arts Press. https://www.gutenberg.org/files/45109/45109-h/45109-h. htm (accessed 4 August 2019).

Erhard, K., Kessler, F., Neumann, N., Ortheil, H.J., and Lotze, M. (2014) Professional training in creative writing is associated with enhanced fronto-striatal activity in a literary text continuation task. *NeuroImage* 100, 15–23.

Faw, B. (2009) Conflicting intuitions may be based on differing abilities: Evidence from mental imaging research. *Journal of Consciousness Studies* 16 (4), 45–68. http://www. ingentaconnect.com/content/imp/jcs/2009/00000016/00000004/art00003 (accessed 22 February 2017).

Federman, R. (1998 [1972]) *Double or Nothing: A Real Fictitious Discourse*. Normal, IL: Fiction Collective Two.

Fernyhough, C. (2010) What do we mean by 'thinking?': Thinking is an active process intimately connected with language. *Psychology Today* (16 August): https://www. psychologytoday.com/au/blog/the-voices-within/201008/what-do-we-mean-thinking (accessed 3 February 2021).

Fernyhough, C. (2014) Top 10 books on memory. *The Guardian* (2 April): https://www. theguardian.com/books/2014/apr/02/top-10-books-memory-charles-fernyhough-pieces-of-light (accessed 27 February 2021).

Fernyhough, C. (2016) *The Voices Within: The History and Science of How We Talk to Ourselves*. London: Profile Books.

Fink, A., Perchtold, C. and Rominger, C. (2018) Creative and cognitive control in the cognitive and affective domains. In R.E. Jung and O. Vartanian (eds) *The Cambridge Handbook of the Neuroscience of Creativity* (pp. 318–332). Cambridge: Cambridge University Press.

Fivush, R. (2011) The development of autobiographical memory. *Annual Review of Psychology* 62, 559–582. https://doi.org/10.1146/annurev.psych.121208.131702.

Fivush, R. (2012) Subjective perspective and personal timeline in the development of autobiographical memory. In D. Berntsen and D.C. Rubin (eds) *Understanding Autobiographical Memory: Theories and Approaches* (pp. 226–245). Cambridge: Cambridge University Press.

Flaherty, A.W. (2005a) *The Mignight Disease: The Drive to Write, Writer's Block, and the Creative Brain*. Boston, MA: Houghton Mifflin.

Flaherty, A.W. (2005b) Frontotemporal and dopaminergic control of idea generation and creative drive. *Journal of Comparative Neurology* 493 (1), 147–153. DOI:10.1002/cne.20768.

Flower, L. and Hayes, J.R. (1981) A cognitive process theory of writing. *College Composition and Communication* 32 (4), 365–387. DOI: 10.2307/356600.

Flying Owl, N. (2013) Ode to an Orange. *PoemHunter.com*: https://www.poemhunter.com/ poem/ode-to-an-orange/ (accessed 18 August 2020).

Foer, J.S. (2006 [2005]) *Extremely Loud & Incredibly Close*. London: Penguin.

Freud, S. (1959 [1908]) Creative writers and day-dreaming. In S. Freud *Collected Papers. Volume 4* (Chapter 9) (J. Riviere, trans.). New York: Basic Books. https://static1.squarespace.com/ static/5441df7ee4b02f59465d2869/t/588e9620e6f2e152d3ebcffc/1485739554918/Freud+- +Creative+Writers+and+Day+Dreaming%281%29.pdf (accessed 11 December 2020).

Furst, G., Ghisletta, P. and Lubart, T. (2017) An experimental study of the creative process in writing. *Psychology of Aesthetics, Creativity, and the Arts* 11 (2), 202–215: http://dx.doi.org/10.1037/aca0000106 (accessed 15 December 2020).

Gardner, J. (1991 [1983]) *The Art of Fiction: Notes on Craft for Young Writers*. New York: Vintage.

Gibbs, R.W. (2011) Metaphors. In M.A. Runco and S.R. Pritzker (eds) *Encyclopedia of Creativity. Volume 2* (2nd edn, pp. 113–119). Amsterdam: Academic Press.

Gilbert, S. (2015 [1930]) Excerpt from *James Joyce's* Ulysses: *A Study*. In *James Joyce Encyclopedia*: http://www.jamesjoyceencyclopedia.com/data/InitialRefs/StuartGilbert/Ulysses%20a%20Study.htm (accessed 7 September 2020).

Ginsberg, A. (2003 [1986]) Cosmopolitan Greetings. *PoemHunter.com* https://www.poemhunter.com/poem/cosmopolitan-greetings/ (accessed 13 March 2019).

Glăveanu V.P. (2018) The cultural basis of the creative process: A dual-movement framework. In T. Lubart (ed.) *The Creative Process: Perspectives from Multiple Domains* (pp. 297–316). London: Palgrave Macmillan.

Glăveanu, V.P. (ed.) (2019) *The Creativity Reader*. New York: Oxford University Press.

Goulbourne, R. (2011) Introduction. In J-J. Rousseau *Reveries of the Solitary Walker* (pp. ix–xxviii) (R. Goulbourne, trans.). Oxford: Oxford University Press.

Greer, G. (2008) The role of the artist's muse. *The Guardian* (2 June): https://www.theguardian.com/artanddesign/artblog/2008/jun/02/theroleoftheartistsmuse (accessed 28 January 2021).

Guerri, G.B. (2018) *Il Vittoriale degli Italiani: Guide*. Milan: Silvana Editoriale.

Heffernan, J.A.W. (2004 [1993]). *Museum of Words: The Poetics of Ekphrasis From Homer to Ashbery*. Chicago: University of Chicago Press.

Heilman, K.M. and Fischler, I.S. (2018) Creativity and the aging brain. In R.E. Jung and O. Vartanian (eds) *The Cambridge Handbook of the Neuroscience of Creativity* (pp. 476–492). Cambridge: Cambridge University Press.

Heilman, K.M., Nadeau, S.E. and Beversdorf, D.O. (2003) Creative innovation: Possible brain mechanisms. *Neurocase* 9 (5), 369–379. DOI: 10.1076/neur.9.5.369.16553.

Henshon, S.E. (2016) A new vision of creativity: An interview with Todd Lubart. *Roeper Review* 38 (1), 3–5. DOI: 10.1080/02783193.2016.1112717.

Herrmann, N. (1997) What is the function of the various brainwaves? *Scientific American* (22 December): https://www.scientificamerican.com/article/what-is-the-function-of-t-1997-12-22/ (accessed 11 January 2021).

Hesiod (2019 [c. 700 BCE]) Hymn 25 to the Muses and Apollo. In *Theogony: The Homeric Hymns* (H.G. Evelyn-White, trans. 1904). http://data.perseus.org/texts/urn:cts:greekLit:tlg0013.tlg025 (accessed 18 February 2021).

Hillis Miller, J. (1995) The ethics of hypertext. *Diacritics* 25 (3), 26–39. https://doi.org/10.2307/465339.

Hippocrates (1994-2009 [400 BCE]) *Aphorisms* (F. Adams, trans. 1891). http://classics.mit.edu/Hippocrates/aphorisms.html (accessed 9 February 2021).

Hogan, P.C. (2014) Literary brains: Neuroscience, criticism, and theory. *Literature Compass* 11 (4), 293–304. https://doi.org/10.1111/lic3.12144.

Holland, N.N. (2009) *Literature and the Brain*. Gainesville, FL: The PsyArt Foundation.

Homer (1888 [c. 800 BCE]) *The Iliad of Homer: With an Interlinear Translation, for the Use of Schools and Private Learners, on the Hamiltonian System* (T. Clark, trans.). Philadelphia, PA: David McKay. https://archive.org/details/iliadhomerwitha00clargoog/page/n12 (accessed 19 November 2019).

Homer (1909 [c. 1200 BCE]) *The Iliad of Homer* (A. Pope, trans., 1715). London: Cassell & Co. https://upload.wikimedia.org/wikipedia/commons/e/e0/The_Iliad_of_Homer. pdf_(accessed 20 July 2020).

Homer (1998 [c. 1200 BCE]) *The Iliad* (R. Fagles, trans., 1990). New York: Penguin.

Horgan, J. (2012) Was James Joyce the greatest mind-scientist ever? *Scientific American* (August 10): https://blogs.scientificamerican.com/cross-check/was-james-joyce-the-greatest-mind-scientist-ever/ (accessed 28 December 2019).

Hughes-Hallett, L. (2013) *The Pike: Gabrielle d'Annunzio: Poet, Seducer and Preacher of War*. London: Fourth Estate.

Iser, W. (1971) Indeterminacy and the reader's response in prose fiction. In J. Hillis Miller (ed.) *Aspects of Narrative: Selected Papers from the English Institute* (pp. 1–46). New York: Columbia University Press.

Iser, W. (1972) The reading process: A phenomenological approach. *New Literary History* 3 (2), 279–299. DOI:10.2307/468316.

Iser, W. (1978) *The Act of Reading: A Theory of Aesthetic Response*. Baltimore, MD: Johns Hopkins University Press.

Jacobs, T. (2017) Debunking myths about creativity and the brain. *Pacific Standard* (14 June): https://psmag.com/news/debunking-myths-about-creativity-and-the-brain (accessed 23 July 2020).

James, H. (1963) The aspern papers [1888]. In H. James *The Turn of the Screw and Other Short Novels* (pp. 153–251). New York: Signet.

James, H. (1972 [1908]) Preface to *The Spoils of Poynton*, 1908. In J.E. Miller, Jr (ed.) *Theory of Fiction: Henry James* (pp. 71–73). Lincoln, NE: University of Nebraska Press.

James, H. (1972 [1911]) Letter to H.G. Wells. In J.E. Miller, Jr (ed.) *Theory of Fiction: Henry James* (p. 77). Lincoln, NE: University of Nebraska Press.

James, H. (2008 [1884]) The Art of Fiction. *Virgil.Org*: http://virgil.org/dswo/courses/novel/james-fiction.pdf (accessed 5 August 2020).

James, W. (1884) On some omissions of introspective psychology. *Mind* 9 (33), 1–26. Available at: www.jstor.org/stable/2246788.

James, W. (1890) *The Principles of Psychology. Volume I*. London: Macmillan. Digitised copy of the 1891 edition available at *Internet Archive*: https://archive.org/details/principlesofpsyc01jameuoft/mode/2up (accessed 29 December 2019).

James, W. (2009 [1880]) Great men, great thoughts, and the environment. Lecture delivered before the Harvard Natural History Society. *Atlantic Monthly* 46 (October), 441–449. Available at: https://www.uky.edu/~eushe2/Pajares/jgreatmen.html (accessed 4 January 2020).

James, W. (2019 [1890]) The stream of thought. In W. James, *The Principles of Psychology. Volume I* (pp. 224–290). London: Macmillan. Available at *Internet Archive*: https://archive.org/details/principlesofpsyc01jameuoft/page/224/mode/2up (accessed 29 December 2019).

Janzer, A.H. (2016) *The Writer's Process: Getting Your Brain in Gear*. Mountain View, CA: Cuesta Park Consulting.

Jennings M.W. (2016) Introduction. In W. Benjamin *One-Way Street* (pp. 1–20) (E. Jephcott, trans.). Cambridge, MA: The Belknap Press of Harvard University Press.

Jewiss V. (2011) Preface. In G. D'Annunzio *Notturno* (pp. vii–xi) (S. Sartarelli, trans.). New Haven, CT: Yale University Press.

Joyce, J. (1966 [1922]) *Ulysses*. London: The Bodley Head.

Jung, C. (1985 [1933]) Psychology and literature (W.S. Dell and C.F. Baynes, trans.). In B. Ghiselin (ed.) *The Creative Process: Reflections on Invention in the Arts and Sciences* (pp. 217–232). Berkeley, CA: University of California Press.

Jung, R.E. and O. Vartanian (eds) (2018) *The Cambridge Handbook of the Neuroscience of Creativity*. Cambridge: Cambridge University Press.

Kandel, E.R. (2000) From nerve cells to cognition: The internal cellular representation required for perception and action. In E.R. Kandel, J.H. Schwartz and T.M. Jessell (eds) *Principles of Neuroscience*. (4th edn, pp. 381–403). New York: McGraw-Hill.

Kandel, E.R., Schwartz, J.H. and Jessell, T.M. (2000) The neurobiology of behaviour. In E.R. Kandel, J.H. Schwartz and T.M. Jessell (eds) *Principles of Neuroscience*. (4th edn, pp. 3–4). New York: McGraw-Hill.

Katz, L. (ed.) (2002–2004) *Classical Monologues. Volumes 1-4*. New York, NY: Applause.

Kaufman, S.B. (2020) The neuroscience of creativity: A Q&A with Anna Abraham: The latest state of the field of the neuroscience of creativity. *Scientific American* (4 January): https://blogs.scientificamerican.com/beautiful-minds/the-neuroscience-of-creativity-a-q-a-with-anna-abraham/ (accessed 8 February 2021).

Kaufman, S.B. (2021) Bio. https://scottbarrykaufman.com/bio/ (accessed 5 February 2021).

Kaufman, S.B. and Kaufman, J.C. (eds) (2009) *The Psychology of Creative Writing*. Cambridge: Cambridge University Press.

Kaufman, S.B. and Gregoire, C. (2016 [2015]) *Wired to Create: Unravelling the Mysteries of the Creative Mind*. New York: TarcherPerigee.

Keats, J. (1884) Isabella; or, The Pot of Basil. *The Poetical Works of John Keats*. London: Macmillan. Available at: *Bartleby.com*: https://www.bartleby.com/126/38.html (accessed 8 February 2021).

Keats, J. (1966 [1818]) Letter to George and Tom Keats. In J. Keats *The Selected Poetry of John Keats* (Paul de Man, ed., pp. 328–329). New York: Signet.

Keats, J. (2020 [1848]) When I Have Fears. Poetry Foundation: https://www.poetryfoundation. org/poems/44488/when-i-have-fears-that-i-may-cease-to-be (accessed 14 August 2020).

Kemp, S. (2018) *Writing the Mind: Representing Consciousness from Proust to the Present*. New York: Routledge.

Kenward, J. (2017) *The Joy of Mindful Writing: Notes to Inspire Creative Awareness*. London: Leaping Hare Press.

King, S. (2000) *On Writing*. London: Hodder & Stoughton.

Kirkham, B.F. [InkdropK] (2014) Ode to an Orange. *All Poetry*: https://allpoetry.com/ poem/11589320-Ode-to-an-Orange--by-InkdropK_(accessed 18 August 2020).

Kluger J. (2019) This is your brain on creativity. *The Science of Creativity* (pp. 11–17). Special TIME Edition. New York: Meredith Corporation.

Koestler, A. (1989 [1964]) *The Act of Creation*. London: Arkana/Penguin Books.

Kozbelt A. (2011) Theories of creativity. In M.A. Runco and S.R. Pritzker (eds) *Encyclopedia of Creativity. Volume 2* (2nd edn, pp. 437–479). Amsterdam: Academic Press.

Krauth, N. (1996) The Swing. In J. Pausacker (ed.) *Hide and Seek* (pp. 134–144). Melbourne: Mandarin.

Krauth, N. (1997) The big theme park: One writer's Queensland. *Australian Book Review* 194, 36–41.

Krauth, N. (2000) A ticket to Albany. *Australian Creative Nonfiction*, The TEXT Special Issue Website Series 1 (April): http://www.textjournal.com.au/speciss/issue1/krauth. htm (accessed 1 February 2021).

Krauth, N. (2006) The domains of the writing process. In N. Krauth and T. Brady (eds) *Creative Writing: Theory Beyond Practice* (pp. 187–196). Teneriffe: Post Pressed. https://research-repository.griffith.edu.au/bitstream/handle/10072/24867/45869_2. pdf;jsessionid=74505FEB80BC84CEB8E16FAB7E6FEC8D?sequence=1 (accessed 1 February 2021).

Krauth, N. (2010) The story in my foot: Writing and the body. *TEXT: Journal of Writing and Writing Courses* 14 (1): http://www.textjournal.com.au/april10/krauth.htm (accessed 4 October 2021).

Krauth, N. (2012) Alcohol and the writing process. *TEXT: Journal of Writing and Writing Courses* 16 (2): http://www.textjournal.com.au/oct12/krauth.htm (accessed 6 August 2019).

Krauth, N. (2015) By the fingernails. *Creative Writing as Research IV,* TEXT Special Issue Number 30, *TEXT: Journal of Writing and Writing Courses* 19 (2), Part I, 88–92. https://textjournal.scholasticahq.com/article/27239-creative-writing-as-research-iv-part-1 (accessed 28 September 2021).

Krauth, N. (2016a) *Creative Writing and the Radical: Teaching and Learning the Fiction of the Future.* Bristol: Multilingual Matters.

Krauth, N. (2016b) Reading the sentence fully: A Shakespeare memoir, *Shakespeare 400,* TEXT Special Issue Number 36, *TEXT: Journal of Writing and Writing Courses* 20 (2): https://doi.org/10.52086/001c.27050 (accessed 28 September 2021).

Krauth, N. (2017) Byron snapshots. *Review of Australian Fiction* 22 (5). http://reviewofaustralianfiction.com/ (accessed 1 February 2021).

Krauth, N. (2019) Fragmented narratives: Minding the textual gap. *TEXT: Journal of Writing and Writing Courses* 23 (2): https://doi.org/10.52086/001c.18601 (accessed 28 September 2021).

Krauth, N. and Bowman, C. (2017) Ekphrasis and the writing process. *New Writing: The International Journal for the Practice and Theory of Creative Writing* 15 (1), 11–30. http://dx.doi.org/10.1080/14790726.2017.1317277.

Kress, G. (2003) *Literacy in the New Media Age.* London: Routledge.

Kress, G. (2010) *Multimodality: A Social Semiotic Approach to Contemporary Communication.* London: Routledge.

Kress, G. (2011) Making meaning: The role of semiotics in education: A conversation with Gunther Kress. London: Institute of Education. http://www.ioe.ac.uk/studentInformation/61188.html (video accessed 11 June 2014).

Lakoff, G. and Turner, M. (1989) *More than Cool Reason: A Field Guide to Poetic Metaphor.* Chicago, IL: University of Chicago Press.

Langer, E.J. (1989) *Mindfulness.* Reading, MA: Addison-Wesley.

Laozi (2006–2019 [6th century BCE]) *The Tao Te Ching* (J. Legge, trans., 1891). Available at Chinese Text Project: https://ctext.org/dao-de-jing (accessed 20 June 2017).

Laozi (2019 [6th century BCE]) Empty space. *New Philosopher* 24, 108–109.

Lawrence, D.H. (1966 [1923]) *Selected Poems.* Harmondsworth: Penguin.

Lawrence, D.H. (1967 [1921]) *Women in Love.* Harmondsworth: Penguin.

Lawrence D.H. (2004 [1927]) Introduction to these paintings. In J.T. Boulton (ed.) *Late Essays and Articles. D. H. Lawrence* (pp. 185–217). Cambridge: Cambridge University Press.

LeDoux, J. (2018) Foreword. In J. Tougaw *The Elusive Brain: Literary Experiments in the Age of Neuroscience* (pp. ix–xii). New Haven, CT: Yale University Press.

Legouis, E. (2000) The influence of Rousseau. In A.W. Ward and A.R. Waller (eds) *The Cambridge History of English and American Literature. Volume XI.* Available at *Bartleby.com: Great Books Online*: https://www.bartleby.com/221/0501.html (accessed 23 December 2020).

Le Guin, U.K. (2004) Stress-rhythm in poetry and prose. In Ursula K. Le Guin, *The Wave in the Mind: Talks and Essays on the Writer, the Reader, and the Imagination* (pp. 70–94). Boston, MA: Shambala.

Leon, D. (2002 [2001]) *A Sea of Troubles.* London: Arrow.

Leonard, E. (1982 [1980]) *Gold Coast*. London: Penguin.

Lesnik-Oberstein, K. (2017) The object of neuroscience and literary studies. *Textual Practice* 31 (7), 1315–1331. https://doi.org/10.1080/0950236X.2016.1237989

Lieberman, M.D. (2013) *Social: Why Our Brains Are Wired to Connect*. Oxford: Oxford University Press. Kindle Edition.

Lindqvist, S. (1996 [1992]) *'Exterminate all the Brutes'* (J. Tate, trans.). New York: The New Press.

Lineen, J. (2019) *Perfect Motion: How Walking Makes Us Wiser*. Melbourne: Ebury/Penguin.

Lineen, J. (2020) Perfect motion: Walking, creativity and writing. PhD submission, Griffith University, Gold Coast, QLD. https://doi.org/10.25904/1912/482.

Lodge, D. (1992) *The Art of Fiction: Illustrated from Classic and Modern Texts*. London: Penguin.

Lodge, D. (2003 [2002]) *Consciousness and the Novel: Connected Essays*. London: Penguin.

Lovett, D. (2017) Siegfried Kracauer and the operative feuilleton. MA dissertation, University of California, Santa Barbara. https://escholarship.org/uc/item/0t72s2bc (accessed 22 February 2019).

Lubart, T. (2009) In search of the writer's creative process. In S.B. Kaufman and J.C. Kaufman (eds) *The Psychology of Creative Writing* (pp. 149–165). Cambridge: Cambridge University Press.

Lubart, T. (ed.) (2018) *The Creative Process: Perspectives from Multiple Domains*. London: Palgrave Macmillan.

MacIntyre, T.E., Igou, E.R., Campbell, M.J., Moran, A.P. and Matthews, J. (2014) Metacognition and action: A new pathway to understanding social and cognitive aspects of expertise in sport. *Frontiers in Psychology* 5 (16 October). https://doi.org/10.3389/fpsyg.2014.01155.

Mann, T. (1992 [1903]) Tristan. In Thomas Mann *Death in Venice and Seven Other Stories* (pp. 363–409) (H.T. Lowe-Porter, trans.). New York: The Modern Library. Available at *Internet Archive*: https://archive.org/stream/deathinvenicesev00mann_0#page/362/mode/2up (accessed 1 February 2021).

Marcus, G. (2016) Preface. In W. Benjamin, *One-Way Street* (pp. ix–xxv) (E. Jephcott, trans.). Cambridge MA: The Belknap Press of Harvard University Press.

Martindale, C. (1995) Creativity and connectionism. In S.M. Smith, T.B. Ward and R.A. Finke (eds) *The Creative Cognition Approach* (pp. 249–268). Cambridge, MA: MIT Press.

McGann, J. (2001) *Radiant Textuality: Literature After the World Wide Web*. New York: Palgrave.

McKiernan, K.A., D'Angelo, B.R., Kaufman J.N. and Binder, J.R. (2006) Interrupting the 'stream of consciousness': An fMRI investigation. *NeuroImage* 29 (4), 1185–1191. https://doi.org/10.1016/j.neuroimage.2005.09.030.

McKinney, M. (2006) Introduction. In Sei Shōnagon *The Pillow Book* (pp. ix–xxix) (M. McKinney ed. and trans.). London: Penguin.

McLoughlin, N. and Brien, D.L. (eds) (2012) Creativity: Cognitive, social and cultural perspectives. *TEXT: Journal of Writing and Writing Courses* 16 (Special Issue Number 13). https://textjournal.scholasticahq.com/issue/3876.

Meeker, M. (1969) *The Structure of Intellect: Its Interpretation and Uses*. Columbus, OH: Charles E Merrill.

Merriam-Webster (2019) 'feuilleton, noun'. *Merriam-Webster Dictionary*: https://www.merriam-webster.com/dictionary/feuilleton (accessed 22 February 2019).

Miller, A. (2015) *Poetry, Photography, Ekphrasis: Lyrical Representations of Photographs From the 19th Century to the Present*. Liverpool: Liverpool University Press.

Milosz, C. (1968 [1959]) *Native Realm: A Search for Self-definition* (C. Leach, trans.). Garden City: Doubleday.

Mooij, J.J.A. (1993) *Fictional Realities: The Uses of Literary Imagination.* Amsterdam: John Benjamins.

Moore, D.W. (2012) *The Mindful Writer: Noble Truths of the Writing Life.* Boston, MA: Wisdom Publications.

Morrell, J.P. (2006) *Between the Lines: Master the Subtle Elements of Fiction Writing.* Cincinnati, OH: Writer's Digest.

Mulrine, A. (1998) Paradise follows beloved and jazz: Toni Morrison talks about her latest novel. *Princeton Alumni Weekly* (11 March), 22.

Nabokov, V. (1989 [1951, 1966]) *Speak Memory: An Autobiography Revisited.* New York: Vintage.

Nagy G. (2018) A re-invocation of the Muse for the Homeric *Iliad. Classical Enquiries* (16 August): https://classical-inquiries.chs.harvard.edu/a-re-invocation-of-the-muse-for-the-homeric-iliad/ (accessed 20 July 2020).

Nęcka, E. (2011) Perception and creativity. In M.A. Runco and S.R. Pritzker (eds) *Encyclopedia of Creativity. Volume 2* (2nd edn, pp. 216–219). Amsterdam: Academic Press.

Neruda, P. (1996 [1954]) Ode to the Orange. *Fifty Odes* (pp. 103–107) (G.D. Schade, trans.). Austin, TX: Host Publications.

Neurohealth (2021) Brain wave frequencies. Neurohealth: https://nhahealth.com/brainwaves-the-language/ (accessed 11 January 2021).

Ong, W.J. (2012 [1982]) *Orality and Literacy: The Technologizing of the Word, with Additional Chapters by John Hartley.* London: Routledge.

Palmer, A. (2004) *Fictional Minds: The Novelist as Voice Hearer.* Lincoln, NE: University of Nebraska Press.

Pessoa, F. (2015 [1982]) *The Book of Disquiet.* London: Penguin.

Piirto, J. (1998 [1992]) *Understanding Those Who Create. Second Edition.* Scottsdale, AZ: Gifted Psychology Press.

Piirto, J. (2002) *'My Teeming Brain': Understanding Creative Writers.* Cresskill, NJ: Hampton Press.

Piirto, J. (2011) Talent and creativity. In M.A. Runco and S.R. Pritzker (eds) *Encyclopedia of Creativity. Volume 2* (2nd edn, pp. 427–434). Amsterdam: Academic Press.

Piirto, J. (2018) The creative process in writers. In T. Lubart (ed.) *The Creative Process: Perspectives from Multiple Domains* (pp. 89–121). London: Palgrave Macmillan.

Plato (2001 [c. 390 BCE]) Ion (P. Woodruffe, trans.). In V.B. Leitch (ed.) *The Norton Anthology of Theory and Criticism (pp. 37–48).* New York: W.W. Norton.

Plato (n.d [c. 370 BCE]) *Phaedrus by Plato* (B. Jowett, trans.). Full Text Archive: https://www.fulltextarchive.com/page/Phaedrus/ (accessed 28 January 2021).

Poe, E.A. (2001 [1846]) The philosophy of composition. In V.B. Leitch (ed.) *The Norton Anthology of Theory and Criticism* (pp. 739–750). New York: W.W. Norton.

Prendergast, J. (2019) Narrative and the unthought known: The immaterial intelligence of form. *TEXT: Journal of Writing and Writing Courses* 23 (1). https://doi.org/10.52086/001c.23539 (accessed 1 November 2021).

Prendergast, J. (2020) Ideasthetic imagining: Patterns and deviations in affective immersion. *New Writing: The International Journal for the Practice and Theory of Creative Writing* 18 (1). https://doi.org/10.1080/14790726.2019.1709508 (accessed 1 November 2021).

Pritzker, S.R. (2012) Substance abuse and creativity. In M.A. Runco and S.R. Pritzker (eds) *Encyclopedia of Creativity. Volume 2* (2nd edn, pp. 390–395). Amsterdam: Academic Press.

Pylyshyn, Z. (2003) *Seeing and Visualizing: It's Not What You Think*. Cambridge, MA: MIT Press.

Rabovsky, M. (2010) Q & A. *Berlin School of Mind and Brain Newsletter No. II: An Issue on Language* (June), 5–17. http://www.mind-and-brain.de/fileadmin/downloads/Ausschreibungen/mab_newsletter_no_02_nachdruck_2_2011_screen.pdf (accessed 5 August 2019).

Ramachandran, V.S. and Hubbard, E.M. (2001) Synaesthesia: A window into perception, thought and language. *Journal of Consciousness Studies* 8 (12), 3–34. http://chip.ucsd.edu/pdf/Synaesthesia%20-%20JCS.pdf (accessed 25 August 2020).

Rapin, R. (1674) *Reflections on Aristotle's Treatise of Poesie*. (T. Rhymer, trans.). London: Printed by Thomas Newcomb for H. Herringman. Available at: https://www.google.com.au/books/edition/Reflections_on_Aristotle_s_Treatise_of_P/hN5oAAAAcAAJ?hl=en&gbpv=1&dq=rene+rapin&printsec=frontcover (accessed 28 January 2021).

Rekulak, J. (2001) *The Writer's Block: 786 Ideas to Jump-start your Imagination*. Philadelphia, PA: Running Press.

Reviews (2020) *Consciousness and the Novel*. Harvard University Press: https://www.hup.harvard.edu/catalog.php?isbn=9780674013773&content=reviews (accessed 5 August 2020).

Rexroth, K. (1957) My head gets tooken apart. *The Nation* (14 December). Available at Kenneth Rexroth Archive, Bureau of Public Secrets: http://www.bopsecrets.org/rexroth/essays/psychology.htm (accessed 15 November 2019).

Reynolds, S. (2015) *Fire Up Your Writing Brain*. Cincinnati, OH: Writer's Digest.

Richardson, D.M. (1921 [1915]) *Pointed Roofs*. London: Duckworth. Available at *Internet Archive*: https://archive.org/details/pointedroofs00richuoft/page/n5/mode/2up (accessed 6 February 2021).

Rousseau, J-J. (2008 [1782]) *Confessions* (A. Scholar, trans.). Oxford: Oxford University Press.

Rousseau, J-J. (2011 [1782]) *Reveries of the Solitary Walker* (R. Goulbourne, trans.). Oxford: Oxford University Press.

Runco, M.A. (2011a) Divergent thinking. In M.A. Runco and S.R. Pritzker (eds) *Encyclopedia of Creativity. Volume 1* (2nd edn, pp. 400–403). Amsterdam: Academic Press.

Runco, M.A. (2011b) Perspectives. In M.A. Runco and S.R. Pritzker (eds) *Encyclopedia of Creativity. Volume 2* (2nd edn, pp. 228–230). Amsterdam: Academic Press.

Russ, S.W. and Dillon, J.A. (2011) Associative theory. In M.A. Runco and S.R. Pritzker (eds) *Encyclopedia of Creativity. Volume 1* (2nd edn, pp. 66–71). Amsterdam: Academic Press.

Sadler-Smith, E. (2015) Wallas' four-stage model of the creative process: More than meets the eye? *Creativity Research Journal* 27 (4), 342–352. DOI: 10.1080/10400419.2015.1087277.

Sagar, K. (2007) *D. H. Lawrence: Poet*. Tirril, Cumbria: Humanities eBooks.

Saporta, M. (2011 [orig. French edition 1962; trans. 1963]) *Composition No 1*. London: Visual Editions.

Sarafoleanu, C., Mella, C., Georgescu, M. and Perederco, C. (2009) The importance of the olfactory sense in the human behavior and evolution. *Journal of Medicine and Life* 2 (2), 196–198. https://www.ncbi.nlm.nih.gov/pmc/articles/PMC3018978/ (accessed 2 February 2021).

Schwartz, B.L. and Perfect, T.J. (2002) Introduction: Toward an applied metacognition. In T.J. Perfect and B.L. Schwartz (eds) *Applied Metacognition* (pp. 1–11). Cambridge: Cambridge University Press.

Sebald, W.G. (2013 [2001]). *Austerlitz*. London: Penguin.

Segal, C. (1995) Spectator and listener. In J.P. Vernant (ed.) *The Greeks* (pp. 184–217). Chicago, IL: University of Chicago Press.

Shah, C., Erhard, K., Ortheil, H.J., Kaza, E., Kessler, C. and Lotze, M. (2011) Neural correlates of creative writing: An fMRI Study. *Human Brain Mapping* 34 (5), 1088–1101. https://doi.org/10.1002/hbm.21493.

Shakespeare, W. (2003–2021 [1595-1596]) *A Midsummer Night's Dream*. Open Source Shakespeare: https://www.opensourceshakespeare.org/views/plays/play_view.php?WorkID=midsummer&Act=5&Scene=1&Scope=scene (accessed 1 February 2021).

Shakespeare, W. (1963 [1600–1601]) *The Tragedy of Hamlet Prince of Denmark*. New York: Signet/New American Library.

Shakespeare, W. (2021 [1599–1600]) *The Life of King Henry the Fifth. The Complete Works of Shakespeare*: http://shakespeare.mit.edu/henryv/full.html (accessed 8 February 2021).

Shelley, M. (2013 [1831]) Introduction. *Frankenstein: Or, The Modern Prometheus. The Project Gutenberg EBook of Frankenstein*, by Mary W. Shelley: https://www.gutenberg.org/files/42324/42324-h/42324-h.htm (accessed 26 January 2022).

Shelley, P.B. (2001 [1840]) A defence of poetry. In V.B. Leitch (ed.) *The Norton Anthology of Theory and Criticism* (pp. 695–717). New York: W.W. Norton.

Shepard, R.N. (1978) The mental image. *American Psychologist* 33 (2), 125–137. DOI: doi.apa.org/journals/amp/33/2/125.pdf.

Shōnagon, S. (2006 [c. 1002]) *The Pillow Book* (M. McKinney ed. and trans.). London: Penguin.

Skov, M., Stjernfelt, F., and Paulson, O.B. (2007) Language and the brain's 'mental cinema'. In B. Grundtvig, M. McLaughlin and L. Waage Petersen (eds) *Image, Eye and Art in Calvino: Writing Visibility* (pp. 184–199). London: Legenda, Modern Humanities Research Association/Maney Publishing.

Smallwood, J. and Schooler, J.W. (2014) The science of mind wandering: Empirically navigating the stream of consciousness. *Annual Review of Psychology* 66, 487–518. DOI: 10.1146/annurev-psych-010814-015331.

Solnit, R. (2001 [2000]) *Wanderlust: A History of Walking*. New York: Penguin.

Steffens, L. (1931) *The Autobiography of Lincoln Steffens*. New York: Harcourt Brace & Co.

Stein, G. (2003 [1912]) Tender Buttons. In G. Stein *Three Lives & Tender Buttons* (pp. 243–300). New York: Signet.

Steiner, G. (1998) Introduction. In W. Benjamin, *The Origin of German Tragic Drama* (J. Osbourne, trans.). London: Verso. http://rebels-library.org/files/benjamin_drama.pdf: 7-26 (accessed 20 November 2015).

Strawson, G. (2002) The mind's I. *The Guardian* (24 November): https://www.theguardian.com/books/2002/nov/23/fiction.highereducation (accessed 5 August 2020).

Suleiman, S.R. (1980) Introduction: Varieties of audience-oriented criticism. In S.R. Suleiman and I. Crosman (eds) *The Reader in the Text: Essays on Audience and Interpretation* (pp. 3–45). Princeton, NJ: Princeton University Press.

Summers, A. (2012) Re: How haiku structures experience. *The Haiku Foundation* (April 13): https://www.thehaikufoundation.org/forum_sm/index.php?topic=2634.15 (accessed 19 January 2020).

Süskind, P. (2006 [1985]) *Perfume: The Story of a Murderer* (J.E. Woods, trans.). London: Penguin.

Takolander, M. (2015) From the 'mad' poet to the 'embodied' poet: Reconceptualising creativity through cognitive science paradigms. *TEXT: Journal of Writing and Writing Courses* 19 (2): https://doi.org/10.52086/001c.25372 (accessed 28 September 2021).

Testard, J. (2012) Interview with Jonathan Safran Foer. *The White Review* 28 (May): https://www.thewhitereview.org/feature/interview-with-jonathan-safran-foer/ (accessed 4 September 2020).

Thoreau, H.D. (1862) Walking. *The Atlantic Monthly* 9 (56), 657–674. https://www.walden.org/wp-content/uploads/2016/03/Walking-1.pdf (accessed 18 February 2021).

Thurston, L. (2008) Splinters of being: Fernando Pessoa as multiple singularity. *Qui Parle* 17 (1), Special Issue: 'Thinking alterity, reprise', 175–192. http://www.jstor.com/stable/20685730 (accessed 28 July 2020).

Tougaw, J. (2018) *The Elusive Brain: Literary Experiments in the Age of Neuroscience.* New Haven, CT: Yale University Press.

Tsur, R. (2007) Issues in literary synaesthesia. *Style* 41 (1), 30–51. https://www.jstor.org/stable/10.5325/style.41.1.30 (accessed 14 August 2020).

Ullmann, S. (1963 [1957]) Panchronistic tendencies in synaesthesia. In S. Ullmann, *The Principles of Semantics* (pp. 266–289). Oxford: Blackwell.

van Gelder, T. (2009) Mindfulness versus metacognition, and critical thinking. Tim Van Gelder (27 May): https://timvangelder.com/2009/05/27/mindfulness-versus-metacognition-and-critical-thinking/ (accessed 13 February 2021).

Vartanian, O., Martindale, C. and Matthews, J. (2009) Divergent thinking ability is related to faster relatedness judgments. *Psychology of Aesthetics, Creativity, and the Arts* 3 (2), 99–103.

Wallas, G. (2014 [1926]) *The Art of Thought.* Tunbridge Wells: Solis Press.

Wartenburger, I. and Spalek, K. (2010) Editorial. *Berlin School of Mind and Brain Newsletter No. II: An Issue on Language* (June), 1-3. http://www.mind-and-brain.de/fileadmin/downloads/Ausschreibungen/mab_newsletter_no_02_nachdruck_2_2011_screen.pdf (accessed 5 August 2019).

Watkins, R. and Krauth, N. (2016) Radicalising the scholarly paper: New forms for the traditional journal article. *TEXT: Journal of Writing and Writing Courses* 20 (1): https://doi.org/10.52086/001c.25303 (accessed 13 February 2021).

Waugh, E. (1930) Gaudi. *The Architectural Review* 67 (403), 309–311. http://search.proquest.com.libraryproxy.griffith.edu.au/trade-journals/gaudi/docview/1459588967/se-2?accountid=14543 (accessed 8 October 2021).

Weir, K. (2019) 'Tasting' yellow, 'hearing' orange. *Brain World* (30 December): https://brainworldmagazine.com/tasting-yellow-hearing-orange/ (accessed 14 August 2020).

Williams, W.E. (1966) Introduction. In D.H. Lawrence *Selected Poems* (pp. 7–9). Harmondsworth: Penguin.

Winokur, J. (ed.) (2000 [1999]) *Advice to Writers: A Compendium of Quotes, Anecdotes, and Writerly Wisdom from a Dazzling Array of Literary Lights.* London: Pavilion.

Woolf, V. (1989 [1929]) *A Room of One's Own.* New York: Harcourt/Harvest.

Woolf, V. (2009 [1919]) Modern fiction. In V Woolf *Selected Essays* (pp. 6–12) (D. Bradshaw ed.). Oxford: Oxford University Press.

Woolf, V. (2011 [1927]) A terribly sensitive mind. In V. Woolf, *The Art of Fiction: A Collection of Essays* (pp. 63–65). Alcester: Read Books. [Originally published in the *New York Herald Tribune*, 18 September 1927.]

Woolf, V. (2011–2021 [1926]) Letter to V. Sackville-West, 16 March 1926. Woolf Online: http://www.woolfonline.com/?node=content/contextual/transcriptions&project=1&parent=48&taxa=49&content=6344&pos=7 (accessed 22 February 2021).

Woolfe, S. (2007) *The Mystery of the Cleaning Lady: A Writer Looks at Creativity and Neuroscience.* Crawley: University of Western Australia Press.

Wordsworth, W. (2002 [1807]) 'I Wandered Lonely as a Cloud.' (From *Poems in Two Volumes*, 1807.) Representative poetry online: https://rpo.library.utoronto.ca/poems/i-wandered-lonely-cloud (accessed 23 December 2020).

Yeats, W.B. (1985 [1912]) Three pieces on the creative process: The thinking of the body. In B. Ghiselin (ed.) *The Creative Process: Reflections on Invention in the Arts and Sciences*. (pp. 106–107). Berkeley, CA: University of California Press.

Yeats, W.B. (1989 [1920]) The Second Coming. *Poetry Foundation*. https://www.poetryfoundation.org/poems/43290/the-second-coming (accessed 4 August 2019).

Zeitlin, F.I. (2013) Figure: ekphrasis. *Greece & Rome* 60 (1), 17–31. doi.10.1017/S0017383512000241.

Zeman, A., Dewar, M. and Della Sala, S. (2015) Lives without imagery: Congenital aphantasia. *Cortex* 73 (December), 378–380. https://doi.org/10.1016/j.cortex.2015.08.015.

Zimmer, C. (2014) This is your brain on writing. *New York Times* (20 June). https://www.nytimes.com/2014/06/19/science/researching-the-brain-of-writers.html__(accessed 5 August 2019).

Index

CPSIA information can be obtained
at www.ICGtesting.com
Printed in the USA
JSHW030804210622
27322JS00005B/29